GOD AND THE BRAIN

God and the Brain

The Rationality of Belief

Kelly James Clark

WILLIAM B. EERDMANS PUBLISHING COMPANY

GRAND RAPIDS, MICHIGAN

Wm. B. Eerdmans Publishing Co.
4035 Park East Court SE, Grand Rapids, Michigan 49546
www.eerdmans.com

Published 2019
Printed in the United States of America

25 24 23 22 21 20 19   1 2 3 4 5 6 7

ISBN 978-0-8028-7691-1

Library of Congress Cataloging-in-Publication Data

A catalog record for this book is available from the Library of Congress.

To my aunts and uncles:
Karen and Kenny Severson,
Ginny and Jim Kelham,
and Tom and Leslie Leech

Contents

Foreword

I have a two-and-a-half-year-old friend who has recently discovered monsters in his bedroom. By normal indicators, he has a *belief* in monsters: in garbled English he can articulate that the monsters are there, he shows emotional signs of genuine fear (and says, "I scary," meaning that he is scared), and he modifies his actions based upon these beliefs. We could easily conclude that this little boy has an irrational belief in monsters—but *is* it irrational? Just because a belief is false does not mean that the belief is irrational. People can sensibly and rationally form beliefs that turn out to be false. Furthermore, do we even know that there are not monsters in his bedroom? Children of his age have many abilities that adults have lost, including the ability to perceive subtle differences in sounds. Maybe preschoolers are also better at monster perception (I hope not). Deciding whether someone else's beliefs are rationally or justifiably formed is no easy business—though it's often tempting to proceed as though it is—and perhaps even more so when the beliefs in question don't match our own.

In today's climate, large and diverse batches of information on people's thoughts, values, and commitments from all over the world are readily available through electronic and print media. This level of accessibility makes two mistakes common when evaluating whether someone's beliefs are virtuously formed and held. The first mistake is to assume that the other's beliefs (such as their religious or moral beliefs) should be mistrusted because of tribalistic sentiment—for example, "those other people haven't been as careful as I have in forming *my* beliefs." In this line of thinking, *my* beliefs are thoughtfully and carefully formed through rational reflection and the reliance on good evidence and sound arguments, and they are backed up by the authority of qualified experts; *their* beliefs are hastily produced through dubious methods. The second mistake is to conclude that the vast diversity in beliefs—about, for instance, whether there are spirits, whether humans have immaterial souls, whether some objects are really sacred, and whether there is a cosmic creator God—implies that no one's beliefs are any better than anyone else's. Who am I to say that my beliefs have been more appropriately formed than anyone else's? The humility in that sentiment is valuable, but it's erroneous to conclude that, on the basis of the vast diversity of beliefs, we can't really know anything or at least be justified in claiming we know it. Both of these mistakes arise from failing to know how beliefs are formed and why it matters.

Enter science.

Ever since psychology began getting its scientific legs beneath it in the late nineteenth century, it has dabbled with trying to account for religious beliefs. The past twenty years, however, have seen an unprecedented blossoming of scientific attempts to explain religious beliefs. Where do they come from? What causes them? Unsurprisingly, such investigations have been quickly followed by asking whether these causal explanations bear on whether religious beliefs are true or false, rational or irrational. Neuroscience, as well as cognitive, developmental, evolutionary, and social psychology, have matured enough in their theories and methods to begin making real headway concerning how beliefs are formed, and this progress has been turned to studying the causes of religious beliefs. Most prominent in these efforts is the interdisciplinary space known as cognitive science of religion, which is the main focus of this book. As in times past, these scientific treatments of religious beliefs have been quickly followed by philosophical treatments concerning what the science means for whether religious beliefs are good or bad, justified or not, rational or irrational. As a philosopher of religion with a gift for making complex ideas and arguments accessible, Kelly Clark has been among the leaders of this philosophical engagement with cognitive science of religion.

Dr. Clark's interest in the area began in the late 1990s when I first became a psychology professor at Calvin College, where he was teaching philosophy. I gave a seminar to my Calvin colleagues, presenting scientific research that would later be known as cognitive science of religion, and Dr. Clark was the only non-psychologist in attendance. When I finished, Dr. Clark made a point of coming over and paying me what I choose to take as a compliment: "Gee, I didn't think you could do anything interesting with psychology." And now here we are a few years later with an entire book on the subject.

Dr. Clark has chosen to write a book that falls at the intersection of philosophy, religious studies, theology, and several scientific disciplines. Doing so is no easy feat, and even in the hands of a communicator as gifted as Dr. Clark, understanding what is at stake is not straightforward. To help a bit, I offer a few distinctions.

First, causes are not reasons. In common talk, we may answer the question, "Why do you think that?" by reference to the *reasons* for belief—for example, "I believe the economy is shaky because of labor-force participation data," or "I think we have enough potatoes for dinner because I counted them and checked the recipe." Reasoning is the process of connecting impressions, feelings, and thoughts to other thoughts. Causes, on the other hand, refer to the mechanisms of belief formation. What are the social dynamics, the psychological processes, the brain states, or the chemical processes that make a belief more likely to come about? An example of this connection might be "I believe a ball is in front of me because light waves reflected off the ball into my eye, causing activation in my retina, which sent a signal to my brain that was processed as indicating an object resembling a ball." But what about a statement like "You believe the Beatles are greater than Drake because you are old"? Traditionally in logic, taking the *causes* for beliefs as invalidating the beliefs is considered a fallacy. Just because your belief is caused in some part by your age group, for instance, does not mean the belief is faulty. Likewise, simply discovering a cause for religious beliefs does not invalidate that religious belief. All beliefs have causes of one sort or another, and the sciences that study belief formation, including cognitive science of religion, are in the business of uncovering these causes. Whether these causes do indeed upset the reasons for belief in some cases is the central question of Dr. Clark's book.

Second, in terms of scientific inquiry and explanation, minds are not brains. Brains, along with the rest of the nervous system, are very important in generating mental states such as thoughts, feelings, and experiences; thus the brain sciences and mind sciences are closely related and even overlap (as in the fields of cognitive neuroscience and neuropsychology). Nevertheless, when it comes to understanding religious or any other sort of beliefs, it is the mind and mental processes that matter more immediately than brains and neural dynamics. Beliefs are mind things; neural activation patterns are brain things. The difference is akin to the difference between learning how a computer works on the level of operating systems (e.g., How do I get this thing to print my document? How do I save this file in another format?) and learning how a computer works at the level of microprocessors and circuit boards (e.g., Why doesn't the sound work anymore? Why is this computer so much faster than that one?).

A third distinction to track when considering whether various religious (or anti-religious) beliefs are virtuously formed and held is the difference between individual- and cultural-level explanations. This distinction is subtle enough that many of us working in the field sometimes lose track of it. (I, too, have been guilty.) Nevertheless, explaining why beliefs are common in a culture versus why an individual holds the beliefs that he or she does are two different things. To illustrate, an individual may have a high commitment to the existence of ghosts because of a frightening personal experience that seemed to be a confrontation with someone recently deceased. Or a cultural group could be characterized as having strong ghost-beliefs, even though very few people in a generation actually witness a ghost, because ghost-beliefs are a strong part of the group's history and are ensconced in regular ritualized practices. Whereas the two levels of analysis are related, an explanation on one level cannot be assumed to simply and completely map onto the other level.

Finally, when considering scientific treatments of religious thought and their implications, I find it helpful to keep in mind that not all "religious" thoughts are formed the same way, even if they carry the label "religion" in popular or scholarly discourse. In fact, from a scientific perspective, "religion" might be no more than a useful heuristic category, much in the way "tree" is only a rough category for tallish, woodier-than-average plants that have appendage-like parts. For a botanist, some things called "trees" are much more closely related to non-trees than to other trees. Likewise, once we start considering the causes for beliefs and practices commonly called "religious," it may turn out that many of these are more closely related to nonreligious beliefs and practices than other "religious" beliefs and practices. I regard belief in gods as having more in common with the "nonreligious" belief that other human beings have minds than with the "religious" belief in karma. The causes for cleansing rituals may be, in some ways, more closely related to how people brush their teeth than to other religious rituals such as weddings. Consequently, a scientific explanation of a particular "religious" belief, such as believing in a cosmic creator God, may have few implications for the status of other "religious" beliefs (e.g., a belief in ancestor spirits) but tremendous implications for the status of some seemingly more distant beliefs (e.g., that living things seem to have value and purpose in the world).

These distinctions, along with the apparent complexities of how beliefs seem to be formed—religious or otherwise—should give us pause before hastily condemning other individuals' or other cultures' beliefs as irrational, unjustified, or unworthy. Finding others' beliefs somewhat mystifying at times is probably inevitable. Nevertheless, as Dr. Clark encourages, when we consider our own and others' beliefs, a scientific consideration of how religious beliefs are formed should encourage a spirit of intellectual humility. My toddler friend may not have monsters under his bed, but I can empathize with him for believing there might be. Hopefully, he'll extend the same grace to me when he's old enough to listen to the Beatles.

JUSTIN L. BARRETT, PhD
Fuller Theological Seminary

Acknowledgments

Chapter 1 borrows from and adapts Kelly James Clark, "Spiritually Wired: The Science of the Mind and the Rationality of Belief and Unbelief," *Revista Brasileira de Filosofia da Religião* 3.1 (2016): 9–35.

Chapter 2 borrows from and adapts Kelly James Clark, "Rappin' Religion's Solution to the Puzzle of Human Cooperation," *Huffington Post*, November 11, 2015; Kelly James Clark, "How Real People Believe: Reason and Belief in God," in *Science and Religion in Dialogue*, 2 vols., ed. Melville Y. Stewart (Malden, MA: Wiley-Blackwell, 2010), 1:481–99; and Kelly James Clark, "Reformed Epistemology and the Cognitive Science of Religion," in Stewart, ed., *Science and Religion in Dialogue*, 1:500–513.

Chapter 3 borrows from and adapts Kelly James Clark, "Is Theism a Scientific Hypothesis? Reply to Maarten Boudry," *Reports of the National Center for Science Education* 35.5, 5.2 (September–October 2015); Kelly James Clark and Justin L. Barrett, "Reidian Religious Epistemology and the Cognitive Science of Religion," *Journal of the American Academy of Religion* 79.3 (2011): 639–75.

Chapter 4 borrows from and adapts Kelly James Clark and Justin Barrett, "Reformed Epistemology and the Cognitive Science of Religion," *Faith and Philosophy* 27.2 (2010): 174–89.

Chapter 6 borrows from and adapts Kelly James Clark, "Atheism, Inference, and Intuition," in *Advances in Religion, Cognitive Science, and Experimental Philosophy*, ed. Helen De Cruz and Ryan Nichols (London: Bloomsbury Academic, 2017), 103–18; and Kelly James Clark, "Is Atheism Irrational?," *Big Questions Online*, https://www.bigquestionsonline.com/2014/01/28/atheism-irrational/.

ACKNOWLEDGMENTS

The appendix borrows from and adapts Kelly James Clark, "Atheism and Inferential Bias," *European Journal for Philosophy of Religion* 9.2 (2017).

Abbreviations

ADD	agency-detecting device
CREDs	credibility-enhancing displays
CSR	cognitive science of religion
eToM	existential theory of mind
HADD	hypersensitive agency-detecting device
HFA	high-functioning autism
NDE	near-death experience
ToM	theory of mind

CHAPTER 1

Disproof of Heaven?

You Just Believe That Because

As a first-year university student, I had a humanities professor who dismissed the whole of Christianity, which he claimed was invented whole cloth by Paul, in a single, sneering, unsubstantiated, anecdotal medical diagnosis. Paul, then known as Saul of Tarsus, converted to Christianity while on a mission to hunt down and imprison or even kill the first Christians; indeed, Saul witnessed and approved of the stoning of Stephen. On his way to Damascus, Saul, "still breathing out murderous threats against the Lord's disciples," is surrounded by a sudden and brilliant flash of light that knocked him to the ground. Then he hears a majestic voice say, "Saul, Saul, why do you persecute me?" When Saul asks the voice to identify itself, he hears this reply: "I am Jesus" (Acts 9:1-5).[1] In a single ecstatic vision, Saul "sees" God, hears and understands that Jesus is God, and learns that Jesus's disciples are God's people. In 2 Corinthians 12:4, perhaps reflecting on this vision, Paul reports that he "was caught up to paradise and heard inexpressible things." One explanation of Paul's ecstatic experience is that God overwhelmed him in spirit and truth. Another explanation, the one offered by my first-year humanities professor, was that Paul, who admitted to having a painful thorn in the flesh, was suffering from temporal lobe epilepsy (TLE); Paul's "visions" were nothing but neural misfirings commonly experienced when undergoing an epileptic seizure. Paul believed that Jesus was God in the flesh because of a complex partial seizure of his temporal lobe.

My professor relied on an easy but dubious way to win an argument: play the "You just believe that because . . ." card. Consider some things that one might hear someone say (or that one might think): "You just believe in raising taxes (in spite of the evidence that wealth redistribution is ineffective) because you are a Socialist." "You just believe that your feelings are important (as opposed to the really good reasons I just gave you) because you are a woman." "You just believe that the earth is only 10,000 years old (in opposition to compelling science) because you are a fundamentalist." By showing the (nonrational or psychological) causes of another's belief ("you believe that *because*"), you think you have undermined or diminished the rationality of the person holding the belief. You condescendingly claim that your opponent (or friend or even spouse, for that matter) rejects obvious evidence because their ideology or gender or religion or prejudices made them do it. *They* are irrational, you triumphantly imply, because their belief was caused by psychological impulses. *You*, on the other hand, are coolly rational, basing your belief on a sober assessment of the compelling evidence. You think that, by revealing their psychological impulses to believe (instead of your rational way), you have shown them to be irrational. You declare yourself the winner.

There are a lot of "You just believe that because . . ." claims in areas related to God and the mind. Most famously, Sigmund Freud argued that religion is a psychologically infantile form of wish-fulfillment: in the face of an uncaring cosmos we feel helpless and guilty, and so we invent a father-like God who grants us security and forgiveness. Freud, in paraphrase: "You just believe in God because you have not grown up and faced reality without your psychological crutch." Contemporary Yale psychologist Paul Bloom offers an explanation of belief in gods based on malfunctioning psychological systems; he goes on to claim that religion is "an incidental by-product of cognitive functioning gone awry."[2] Biologist Richard Dawkins similarly argues that the irrationality of religion is a by-product of a built-in irrationality mechanism in the brain. You can hear Bloom and Dawkins saying, "You just believe in God because of a malfunctioning cognitive faculty." We will examine a related claim that God is nothing but a brain spasm because religious experiences are simply neural processes in the brain—"You just believe in God because the neurons in your brain's temporal lobe were overstimulated"—and a claim based on the so-called God gene, which alleges that some humans are, and others aren't, genetically disposed to spiritual beliefs: "You just believe in God because your genes predisposed you to believe." Uncover the neurological, psychological, or genetic substrata of a belief, so the claim goes, and you have thereby undermined it.

Although we will discuss in some detail both of these claims, first we will examine near-death experiences (in which possibly dead individuals claim to have experiences of God): "You just believe that because your stressed-out, nearly dead brain was awash in chemicals." When we separate out some of the hysterical chaff, we will focus on the well-established scientific wheat. And then, in the remainder of the book, we can ask if that wheat, properly understood, undermines rational religious belief.

Proof of Heaven

In 2008, rare and deadly bacteria began feasting on the brain of Harvard neurosurgeon Eben Alexander. While he valiantly resisted the hidden invaders, slowly but surely the bacteria overwhelmed. Alexander's brain eventually succumbed, and he slipped into a deep coma. For seven days he was constantly monitored by his physicians, who clinically documented his decline: his neocortex, the part of the brain that most clearly makes us human, had completely shut down. His rector and friend, Rev. Michael R. Sullivan, was called to his side and prepared to read him his last rites. Just as Alexander's doctors were on the verge of shutting off his life-support system, he sprang back to consciousness.[3]

Before entering the hospital, Alexander did not believe in God. "No scientific proof," he said. When Alexander woke up from the coma, he was a convinced believer in God and the afterlife.

His account was published in a cover story in *Newsweek*. During his coma, while his brain was turned off, Alexander's consciousness left his body, or so he claimed, and traveled into an inexplicably beautiful world, guided by a startlingly beautiful woman. His consciousness, freed from his brain, wandered freely through a muddy darkness and into an embracing light. Here is what he experienced:

I was in a place of clouds.

Big, puffy, pink-white ones that showed up sharply against the deep blue-black sky.

Higher than the clouds—immeasurably higher—flocks of transparent orbs, shimmering beings arced across the sky, leaving long, streamer-like lines behind them.

Birds? Angels? These words registered when I was writing down my recollections. But neither of these words do justice to the beings themselves, which were quite simply different from anything I have known on this planet. They were more advanced. *Higher*.

A sound, huge and booming like a glorious chant, came down from above, and I wondered if the winged beings were producing it. Again thinking about it later, it occurred to me that the joy of these creatures, as they soared along, was such that they *had* to make this noise—that if the joy didn't come out of them this way then they would simply not otherwise be able to contain it. The sound was palpable and almost material, like a rain that you can feel on your skin but that doesn't get you wet.

Seeing and hearing were not separate in this place where I now was. I could *hear* the visual beauty of the silvery bodies of those scintillating beings above, and I could see the surging, joyful perfection of what they sang.[4]

At the end of his journey, his lovely guide spoke to him without sound and without words. She said, "You are loved, deeply cherished, forever. There is nothing you have to fear. You will always be loved, and there is nothing that you can do wrong."[5]

Finally, she told him that he had to return to this world, to his life. Then he woke up.

Alexander had the richest, most real experience of his life at precisely that time when the part of his brain involved in consciousness, thought, memory, and emotion was completely turned off. His deepest thoughts and most profound emotions, which would coalesce into his deepest memory, occurred without the support of his brain. He journeyed in brain-free consciousness into a newer, larger, better world and, so he claims, experienced God's love face-to-face.

Prior to his own experiences of the next world, he had always poohpoohed claims of out-of-body experiences, believing them to be scientifically explicable—perhaps near death the brain is flooded with neurochemicals that produce these remarkable sensations.

But, Alexander wondered, how could he have had such experiences when the neural superstructure of such experiences had completely collapsed? He didn't have a *place* within his body to produce such experiences.

And so Alexander came to believe—was forced, really, to believe—in the eternity of our souls, that there is a bigger and better and longer life after this life, and that God is waiting to embrace us.

After interviewing Alexander, Oprah Winfrey exclaimed, "I just talked to the man who saw God."

There were, of course, the predictable skeptical and sarcastic responses to Alexander's "proof of heaven." Max Read, editor and blogger, said that this is "possibly the most embarrassing cover story *Newsweek* has ever run." He proceeded to deconstruct, line by line, Alexander's account of heaven by comparing it with firsthand accounts of tripping on LSD or mushrooms.[6] Sam Harris, famed atheist, wrote: "Alexander's account is so bad—his reasoning so lazy and tendentious—that it would be beneath notice if not for the fact that it currently disgraces the cover of a major newsmagazine." Alexander's conversion, he mocks, "required a ride on a psychedelic butterfly."[7] Physicist Victor Stenger, author of *God: The Failed Hypothesis*, dismissed Alexander's account as ignorance pure and simple.[8]

Oliver Sacks, professor of neurology and renowned author, joined the skeptics. Such out-of-body experiences, he argued, are illusions that prey on precisely the same portions of the brain that process and store very real experiences.[9] They seem real because they occur in the real-experience portion of the brain and are stored in the real-memory portion of the brain. Such illusions, then, have the inescapable feel of reality; such memories of a spirit world are indistinguishable from memories of a long-ago trip to Disneyworld. Alexander's illusions were nothing but neurological events in his poorly functioning brain.

The pre-coma Alexander may have written something similar.

Hallucinations

Oliver Sacks has his own story of an unusual contact. Sacks was once hiking alone in the mountains of Norway when he happened upon an enormous and cantankerous bull. The bull startled him, and as he fled, he fell down a steep cliff, landing with his leg twisted beneath him. With his dislocated knee in excruciating pain, he fashioned a splint from his umbrella and anorak and began his lonely and painful descent. On the way, believing himself to be near death, he began feeling helpless and increasingly desperate. His body was screaming "Give up," and his mind was beginning to agree. He was just about to stop and rest when he heard "a strong, clear, commanding voice, which said, 'You cannot rest here—you cannot rest anywhere. You have got to go on. Find a pace you can keep up and go on steadily.'" Yielding to the voice, he found the strength to carry on in spite of the crippling pain in his useless leg. He writes, "This good voice, this Life voice, braced and resolved me. I stopped trembling and did not falter again."

Where some might have come to believe that they had heard the still, small voice of God and given thanks, Sacks, instead, claims the voice was a hallucination. He attributes his hallucination to perfectly ordinary and not uncommon cognitive processes.

But suppose it was not a hallucination.

If there is a God, one who occasionally speaks to people in dire circumstances like those Sacks found himself in, there is a not implausible scientific explanation for Sacks's unbelief. Autistic individuals, studies suggest, lack the mental tools necessary for relating to a personal God. (I discuss the relation between autism and unbelief in chapter 7.) And Sacks may have suffered from a mild form of autism. Autistic individuals, who may experience very mild to severe forms of autism, lack, to various degrees, the ability to impute thoughts, feelings, and desires to personal agents. This undergirds, again to various degrees, difficulties in feeling or expressing empathy, which can hinder their ability to enter into normal interpersonal relationships. The loving parent may speak to them, reach out to them, and embrace them, but some autistic children may be incapable of recognizing and responding to them.

In short, autistic individuals have difficulties cognizing a personal relationship with God (if there is a God). God may speak to them, reach out to them, and embrace them. But they find it difficult to recognize a personal God.

Autism has various symptoms, in varying degrees. Most notable, of course, are difficulties with social interactions. The autistic individual's inability to start or maintain conversations may, in children, be attributed to shyness. They are often loners, unable to make friends or sustain friendships, preferring to spend time alone. But in the autistic child, the diagnosis of shyness gives way to a much more pervasive and persistent condition.

Is it possible that Sacks was afflicted with such a condition? Sacks suffered his entire life from a malady called prosopagnosia, more popularly called "face blindness." Those who have face blindness lack the ability to recognize or remember faces, sometimes even the faces of members of their own family or close friends. Sometimes when they look at their mother, for example, they may see a stranger; they may recognize their mother's smell or distinctive gait, but they simply cannot recognize their mother's, or anyone else's, kindly gaze. Sacks was often incapable of recognizing his own face in a mirror. Interesting as this may be, here is the key point for our discussion: face blindness is common among people with autism spectrum disorders. How can you tell what a person is thinking or feeling when you cannot distinguish a face from an object, recognize the person speaking to you, or "read" a face? And if you cannot tell what a person is thinking or feeling, you cannot respond to them as persons.

Sacks has described himself as mostly a loner and his shyness as a "disease." Given his face blindness, difficulties in communication, and lonerism, one might (as an amateur psychologist) conclude that Sacks is a high-functioning autistic. If you were to make such a diagnosis, you would join company with some of Sacks's friends, some of whom are trained psychologists. In an interview, Sacks states: "I was and remain somewhat shy. I don't readily open conversations; I certainly think difficulty recognizing people plays a part there. I have been said to suffer social phobia [or] Asperger's [but I think] that overstates it."[10]

I am not a psychologist, and my point is not to diagnose Sacks. My point is this: if Sacks were autistic, he might be incapable of responding to a personal God.

This is the science speaking now, not the drugs. Autistic individuals, to varying degrees, find it difficult to grasp the personal clues—verbal and nonverbal expressions of thoughts, feelings, and desires—that are essential to personal relationships; so, if there is a God and he is speaking, autistic individuals may be unable to understand and respond.

Suppose that God had, in fact, spoken in his still, small voice to Oliver Sacks. If Sacks is autistic, he might have been constitutionally incapable of recognizing and responding to God's voice.

Near-Death Experiences

So-called near-death experiences (NDEs) have occurred around the world and throughout human history. Alexander's account fits a standard pattern that typically includes reports of pure bright light, floating out of one's body, and a journey often through a tunnel into another dimension (usually, but not always, heaven). Along the way, one often meets spiritual beings who guide one into the next world and, after conducting a review of one's entire life and then securing one's forgiveness, guide one back again into one's body. NDEs seem more real than this-worldly experiences; our world is but a faint shadow of this bigger, better, more substantial, "realer than real" world to come. The colors and flavors and feels of the afterworld are vastly more vivid and powerful and enticing than those in one's relatively flat and dingy and bland earthly life. And so one returns to one's body in the mundane world only very reluctantly, often with a sense of duty after having been instructed to return and tell what one has seen and heard and felt—heaven, usually, and God and love. That, give or take, is the long or short of the typical NDE.

The standard skeptical account of Alexander-like near-death experiences holds that such vivid NDEs are the last gasps of a dying brain on earth, not the first gasps of a new life in heaven. That sense of peace and tranquility, the sensation of passing through a tunnel toward the light, and the feelings of warmth and even of being unconditionally loved are, so this account goes, nothing more than the afterglow of a super-stimulated brain awash in chemicals. Due to the stress of dying, the brain goes into hyper mode— with neurons prodigiously firing and chemicals releasing everywhere. Loss of blood flow to the brain might narrow visual sensations, thus explaining the "perceptions" of a tunnel. The brain is bathed in endorphins, the body's naturally produced morphine, creating a euphoric high accompanied by that peaceful, easy feelin'.

NDEs prompted a recent study that examined the EEGs (electroencephalograms) of rats that had been forced into cardiac arrest. For the first thirty seconds after "death," the EEGs showed "a transient and global surge of synchronized gamma oscillations, which display high levels of inter-regional coherence and feedback connectivity as well as cross-frequency coupling with both theta and alpha waves," which, translated into lay-speak, means that a whole lotta stuff was going on inside those little but "highly aroused" rat brains. Did the EEGs record the brain activity of those little vermin as they excitedly crossed over into the mammalian promised land, forgiven finally for spreading the plague and being disproportionately terrifying to human beings? Or did they show, once and for all, that while NDEs may seem realer than real, they are nothing but doped-up illusions? Did the rat experiments induce a rat high or did they convey their subjects to the basement of rat heaven? The experimenters' sober conclusion: "By presenting evidence of highly organized brain activity and neurophysiologic features consistent with conscious processing at near-death, we now provide a scientific framework to begin to explain the highly lucid and realer-than-real mental experiences reported by near-death survivors."[11] Again, translated into lay-speak: "We don't have any idea what the rats were thinking, sensing, or feeling during their NDEs or how their NDEs relate to human NDEs (how could we, really?), but wow, maybe we can begin to explain away the illusory experience of survivors of NDEs." Or something like that.

Alexander and other survivors of NDEs need not deny the transient and global surge of synchronized gamma oscillations. If the mind/brain is involved in cognizing the NDE, which is by all accounts mind-blowingly awesome and life transforming, one might expect EEG readings that are off the charts. If a person were, for the first time, having to wrap her ordinary mind around an extraordinary Reality, one that is realer than real, neuroloscientists (I made this term up) might see "high levels of interregional coherence and feedback connectivity" as well as "cross-frequency coupling with both theta and alpha waves" (heck, maybe a few other waves as well). Given that we think with our mind/brain, any human experience would be mediated by mind. So, if one were to cross over to the other, awesome side, the brain would be in frantic overdrive trying to grasp it.

Are NDEs proof of heaven, as Alexander claims, or are they all in the mind, as his detractors claim? Did Alexander touch the Really Real, or was it, as Sacks claims, a profound hallucination? So far, the science seems disappointingly neutral. And yet these questions raise this important issue: Are our God-beliefs merely brain events, or do they, at least in certain circumstances for some people, put us in touch with a reality outside of our minds?

The God Helmet

Thank God you don't have to die to touch the sky. Near death is not the only way people have experiences of God; there are other, considerably less dangerous options. Consider the so-called God helmet developed by neuroscientist Michael Persinger. A pilgrimage to Persinger's laboratory at Canada's Laurentian University might culminate in a vision of God. "Vision of God" overstates things a bit. In fact, lots of claims based on the God helmet are exaggerated. We will come back to that. The God helmet, so it is claimed, artificially induces an experience of God by electromagnetically stimulating the brain (with no known involvement on the part of the Almighty). Persinger alleges that the God helmet shows that all claims to have experienced God are the effects of electromagnetic stimulations in the brain. "God" is nothing but the result of transient electrical massages of the brain. As Persinger puts it, "Instead of God creating our brains, our brains created God."[12]

In Persinger's laboratory, white-coated technicians escort each subject through a massive steel door and into a sterile acoustic chamber. They are seated in a comfortable chair and then told they are participating in a relaxation study. The technicians snugly fit a motorcycle helmet, outfitted with electromagnetic solenoids, on their heads. The techies exit, leaving the subject sealed alone in a completely silent, completely dark chamber. Electricity begins flowing into the helmet's four magnetic coils on each side of the head, passing through the electrodes attached to the subject's temples and into their temporal lobe, electro-massaging a small portion of the subject's brain.

The scientists monitor and record brain waves, but the good stuff is reserved for the subject. When the subject's temporal lobes are electromagnetically stimulated, they report various spiritual experiences, including sensing the felt presence of God (or other spiritual beings such as angels, ancestors, and ghosts). Up to about 80 percent of Persinger's subjects reported a "felt presence." Some reported feeling a sense of cosmic harmony or oneness.

Has Persinger identified, isolated, and stimulated the neural "God spot," that part of the brain that creates *ex nihilo* God-beliefs?

Imagine there's no heaven (it's easy if you try), but everyone can have a God helmet. You can purchase your own commercially produced God helmet online for just $145 plus $5 shipping (USA); you can even download plans and make your own. With the God helmet, one gets the goods—inner peace, tranquility, and a sense of harmony—but without religion's cost, such as dull and time-consuming worship services, expensive tithing, tedious prayer, and demanding fasting (not to mention, among many other extreme rituals, massive scarification, fire walking, multiple skin piercings, teeth chiseling, and flagellation). One gets God, so to speak, without all the religious costs (nothing to kill or die for, to follow our John Lennon theme). And with that deep and abiding sense of harmony and oneness, we might imagine, adapting a line from Lennon: "I hope someday you'll join us [owning a God helmet]/ And the world will live as one."

Cue God Helmet Altar Call

Yet maybe not. Let me proceed first by way of a story. When I was in college, there was (more than once) a raucous party across the hall from my apartment. Students from all over campus would bring every variety of liquor, which was then poured, along with everyone else's contribution, into a trash can, mixed together, and then imbibed in copious quantities (with the expected effects). One evening, a few hours into the party, a Moonie (a devotee of the Reverend Sun Myung Moon, infamous founder of the Unification Church) knocked warily upon the door of the boisterous revelers. When her knocks could not be heard above the loud music, she banged again, harder. Finally, someone opened the door and invited her into the party (asking for her contribution to the liquid refreshment). She replied, meekly and quietly, "I am not here for a party. I am selling candles to raise funds for the Unification Church." When the people inside asked what the unduly well-dressed and overly sober young woman wanted, the man at the door yelled out over the music, "Hey everybody, this girl is selling camels for God."

There is the part that really happened—girl selling candles for her church—and then there are the exaggerated reports of what happened—girl selling camels for God. The God helmet is like that.

First things first: the God helmet does not work for everyone and maybe even for no one. While Persinger claims it worked for 80 percent of his subjects, a Swedish lab was unable to replicate his results.[13] The lead scientist, Pehr Granqvist of Uppsala University, attributed Persinger's astonishing "success rate" to suggestibility, and in two ways. First, he claimed that either Persinger or his technicians created in certain suggestible subjects an expectation of a spiritual experience (suggestibility is part and parcel of our native desire to please—manifested in our unconscious eagerness to perform as expected). Second, Granqvist claimed that Persinger's leading questions elicited hoped-for responses concerning felt presences. But in Granqvist's lab, double-blind ruled the day: neither his subjects nor his technicians who worked with them were aware of the purpose of the study; the subjects could not have been susceptible to and the technicians could not have provided subtle clues to the study's purposes. The Granqvist study showed no appreciable spiritual effects for participants who were electromagnetically stimulated. Astonishingly, fully half of the subjects in Granqvist's control group (this group wore a God helmet but received no electromagnetic stimulation whatsoever) reported strong religious experiences! Finally, Granqvist argued that the electromagnetic fields involved in Persinger's experiments, with magnetic fields weaker than those of a refrigerator magnet, were too weak to have any meaningful effect on the brain.

Granqvist's study has led to rounds of increasingly strident responses. Persinger insisted on his scientific bona fides ("It was, too, double-blind"), claiming that Granqvist did not set up his experiment properly. Granqvist and his team demurred. We will leave it at that.

Richard Dawkins, famed atheist, made his own pilgrimage to Persinger's lab to feel what it is like to be a religious believer. Despite Persinger's best intentions and efforts, Dawkins neither saw nor felt anything remotely spiritual. In the 2005 BBC documentary *God on the Brain*, a disappointed Dawkins recalls: "It pretty much felt as though I was in total darkness, with a helmet on my head and pleasantly relaxed." And nothing else—no sensed presence and no sense of oneness with reality (and certainly no vision of God). Persinger alleges that Dawkins's negative result was due to a bout of heavy drinking just prior to entering his laboratory. He need not have been so defensive: one recalcitrant subject does not a refutation make.

Suppose, for the sake of discussion, that Persinger is right and Granqvist and Dawkins are wrong: the God helmet works.

Setting aside problems of replication and suggestibility, Persinger's results are less sensational and stunning than one might expect. While 80 percent felt a personal presence in the room (God, ghosts, ancestors, other people, etc.), did anyone really "see" God? Unlike Dawkins, psychologist Susan Blackmore reported a decidedly different God helmet experience; she reported that it was one of "the most extraordinary experiences I have ever had."[14] But a profound, positive, and even extraordinary experience is not the beatific vision. In fact, only about 1 percent of those involved in Persinger's studies claim to have felt the presence of God. One might expect more than 1 percent to randomly feel the presence of God simply by sitting in a soundless, pitch-black room while participating in a relaxation study (even more to feel a "sensed presence").

Finally, suppose that Persinger has actually succeeded in occasionally inducing visions of God through the electromagnetic stimulation of the brain. Should that undermine rational belief in God?

Given that we are embodied creatures, ones for whom thinking is mediated by brains, we should not be surprised to learn that the brain is deeply implicated in every sort of belief, including God-beliefs. Moreover, religious believers with a strong sense of creation should willingly concede the goodness of the body and, thus, of embodied cognition (which holds that we think with our whole bodies, which include brains). We aren't ghost-like spirits who float above physical reality, forming beliefs without the influences of the body. As creatures of the dust, we are part and parcel of the physical world, and so we process our experiences through our very physical brains. We may be more than brains, but we have brains, and our very human, creaturely cognition is deeply tied to our brain's neural processes. As with, say, perceptual or memory beliefs, particular portions of that brain are much more likely than others to be implicated in or to mediate God-beliefs. Perhaps under very special circumstances, those portions of the brain, if stimulated appropriately, can even generate God-beliefs.

Most of our embodied neural processors are involved in very ordinary cognitions. Consider those involved in perceptual cognition. When I look at a tree and form the belief "There is a tree before me," I do so partly because some portion(s) of the brain are involved in perception (and partly because there is a tree out there, which I see). That is how ordinary perception works—when I see a tree (a real tree, out there, outside my mind), it causes visual information to pass through my eyes by way of my retinal nerves into the visual portion of my brain, which induces various chemical and neural processes, which in turn generate a sensation of a tree and an accompanying tree-belief. Of course, we know very little about how all of this works. But we do know that the perception of a tree involves trees, eyes, nerves, the brain, chemicals, neurons and neural processes, inner sensations, and beliefs.

Perhaps one day a clever neuroscientist will succeed in electromagnetically inducing in the perceptual portion of my brain, without my seeing a tree, a very real visual sensation of a tree. If so, I would find myself believing (wrongly), "There is a tree before me."

Should this electromagnetic creation of a perceptual sensation and corresponding belief undermine the rationality of all of my perceptual beliefs?

I think it is clear that on this particular occasion my perceptual belief, "There is a tree before me," is not rational because it was produced by the direct electromagnetic stimulation of my brain (and not by seeing a tree). The neuroscientist and I would have a good laugh together with me marveling that a machine could so effectively and powerfully reproduce the sensations that I have when I actually perceive something. And I would understand somewhat better the fact that my perceptual encounters with the outside world electromagnetically stimulate portions of my brain in ways that produce inner sensations and even beliefs.

Should I also conclude, from that electromagnetically induced illusion, that all of my perceptual beliefs are illusory?

I think not. While I might be surprised that under very special laboratory circumstances perceptual sensations and beliefs can be induced in me without instigation from the physical world (say, a tree or a dog), I shouldn't think that this entails that all of my perceptual beliefs have been induced in me without instigation from the physical world. Why would I think that? Why suppose that I should stop trusting my perceptual faculties?

Likewise, I will not stop trusting my memory faculty even if a clever neuroscientist should successfully induce in me a memory, or my moral faculty even if a neuroscientist should successfully induce in me a moral belief. And I will not stop believing that my wife loves me even if a clever neuroscientist should induce in me both a very real sense of other persons or of being loved. And it should not trouble me if I were to find out that clever neuroscientists had done any of that in other persons, even lots of them.

How might such neuroscientific findings undermine rational God-beliefs? Not through the discovery that in very unusual circumstances, those involving the direct stimulation of a person's brain, one can induce God-beliefs. Of course, a God-belief produced in that very unnatural way, without any outside connection to the external reality it affirms, would not be rational. But then neither would thusly produced tree-beliefs or memory beliefs. Consider the tree-belief: if I know that my tree-belief was produced directly through the electromagnetic stimulation of my brain (thus not by seeing a tree), then that belief is irrational. That particular tree-belief is rationally undermined by my awareness that it was directly caused by the helmet and not by seeing a tree. Likewise, if I know that my God-belief was electromagnetically induced without any reliable connection to God, then that belief is irrational. That particular God-belief is rationally undermined by my awareness that it was caused by a helmet and not by God.

But does that undermine every tree- or God-belief? Why should such laboratory experiences undermine the rationality of every tree- or God-belief? What about my prior God-beliefs? And what about everyone else's God-beliefs? Would my and a few others' electromagnetically induced beliefs undermine their rationality with respect to their God-beliefs?

To undermine the rationality of all religious believers, we would need some reason to think that most religious beliefs were produced by something like direct electromagnetic stimulations of the brain (not, ultimately, caused by God). Maybe they are—perhaps regular shifts in the earth's tectonic plates have caused electromagnetic eruptions sufficient to induce in large numbers of people various God-beliefs throughout most of the world and through human history. Maybe atheism is on the recent rise because the earth's crust has sufficiently stabilized so that it emits fewer electromagnetic stimulations to the God part of the brain.

"Perhaps" and "maybe" and "if," though, aren't good science. Unless and until someone has shown that most God-beliefs are electromagnetically produced without recourse to God, a few extraordinarily induced God-beliefs aren't sufficient to undermine the rationality of every religious believer.

We are very far from perception helmets, memory helmets, and moral helmets. We are probably even further from other-person helmets and feeling-loved helmets. As of yet, there is no reason to believe in a God helmet (one that induces, beyond randomness, a sense of God). At best, Persinger has invented an occasionally-awesome-feeling-that-sometimes-involves-a-sense-of-others helmet. And even if any of the above were to be invented, they would undermine just one's immediately produced belief, not trust in the cognitive faculties that are typically involved in their production. In order to cease trusting perception or memory, say, one would need good reason to think that most perceptual or memory beliefs were produced by direct electromagnetic stimulation and not by, say, a tree or an event from one's past.

And in order to cease trusting one's God-faculty, one would need good reason to think that most God-beliefs were produced by something like direct electromagnetic stimulation and not by God. Until that has been shown, neuroscientific claims to have undermined the rationality of belief in God lack sufficient support.

The God Gene

In 2004, Harvard-educated molecular biologist Dean Hamer published his sensationally titled book *The God Gene: How Faith Is Hardwired into Our Genes.*[15] The book was featured on the November 29, 2004, cover of *Time* magazine with the provocative subheading "Does our DNA compel us to seek a higher power? Believe it or not, some scientists say yes." In his book, Hamer claims that he has located the gene, the God gene, responsible for human spirituality (the VMAT2 gene). The God gene, he claims, codes for the release of certain intoxicating brain chemicals that, when released, produce spiritual feelings. In the article he says, "I am a believer that every thought we think and every feeling we feel is the result of activity in the brain. I think we follow the basic law of nature, which is that we are a bunch of chemical reactions running around in a bag."[16] God is all in our genes.

In his study, Hamer assessed his subjects' religiosity using the "self-transcendence" portion of the temperament and character inventory (TCI), which measures spirituality, vaguely understood as considering oneself an integral part of the universe. While his scale did not assess belief in a higher being, it did allow assessments of self-forgetfulness (one's ability to be immersed in the moment), harmony (identification of oneself as a part of the universe as a whole), and mysticism (one's degree of openness to the unexplained). Some sample true-false questions that aim at understanding one's level of self-transcendence include:

1. I often become so fascinated with what I am doing that I get lost in the moment—like I am detached from time and place.
2. I often feel so connected to the people around me that it is like there is no separation between us.
3. I am fascinated by the many things in life that cannot be scientifically explained.
4. Often I have unexpected flashes of insight or understanding while relaxing.
5. I sometimes feel so connected to nature that everything seems to be part of one living organism.

Hamer next argued that various mood-regulating chemicals in the brain, monoamines (including serotonin and dopamine), are responsible for the positive and sometimes euphoric feelings of self-transcendence that he associates with spirituality.

Supposing monoamines as key to the chemical reactions running around in the human bag that are positively associated with spirituality, his next step was to find the genetic basis of the production of monoamines (no easy task given that humans have 25,000 protein-coding genes).

Hamer examined his subjects' DNA samples in search of the genes that produced self-transcendence in people. He lighted on the gene known as VMA T2, which is involved in coding for proteins that make up monoamines. Since monoamines are positively correlated with self-transcendence, he believed he had found the self-transcendence gene. And, after he sent the book to a publisher, the self-transcendence gene became the God gene. God is in our genes. According to Hamer, this, and not a faithful and free response to the Transcendent, is why people believe in God.

Such sensational claims smack of genetic determinism—the claim that, just as we have genes that determine the shape of our noses, say, or baldness, so, too, we have genes that determine every aspect of human behavior from being a loner to having a vicious temper. The murderer might claim that he was genetically disposed to violence (and so couldn't help himself). A recent study discussed whether there is a gene for thrill-seeking. We read (and often believe), "There is a gene for that," a gene for, say, generosity, shyness, courage, or compassion. But such claims are completely unsupported by science and ignore the significant role that the environment plays in influencing human behavior. Moreover, genetic determinism violates our deepest sense of human dignity and free will. Our genes and environment may incline us toward certain behaviors and beliefs, but they don't compel or cause those behaviors and beliefs. At least there is no scientific reason to think they do.

Hamer had previously made a rushed, oversimplified, and unsubstantiated claim about genes and behavior. In 1993 he sensationally reported a genetic link to male homosexuality in a region of the X chromosome. Although Hamer would temper his claim to 99.5 percent certainty with "suggest," "seems to indicate," and "probably," the media incautiously rushed in, claiming that science had discovered the gay gene (not entirely without some serious but misleading support from Hamer's study). No other researchers could replicate his results.

But his own personal cautionary tale didn't stop Hamer from rushing *The God Gene* into publication without sufficient nuance or qualification.

The argument of the God gene was based on an ambiguous, unreplicated, and unrefereed study. Even if substantiated, it measured a slight genetic tendency toward a very vague spirituality—unity, harmony, and mysticism. It said little or nothing, because people weren't assessed, about belief in God. Carl Zimmer, in his blistering review of Hamer's book in *Scientific American*, suggested changing the title to *A Gene That Accounts for Less Than One Percent of the Variance Found in Scores on Psychological Questionnaires Designed to Measure a Factor Called Self-Transcendence, Which Can Signify Everything from Belonging to the Green Party to Believing in ESP, According to One Unpublished, Unreplicated Study.*[17]

There is simply no reason to believe that "You just believe in God because your genes made you do it."

Reality or Delusion?

The God gene, the God helmet, and scientific "explanations" of near-death experiences are the most controversial and sensational ways of explaining God away based on the science of the brain. In every case the science is unsettled, and yet it is often asserted as settled (with allegedly disastrous consequences for rational belief in God). There is, however, a firmly established and widely held science of the brain and God—the cognitive science of religion, the science of the mind in relation to religious beliefs and practices. Cognitive science suggests that the brain naturally disposes or inclines people toward God-beliefs. Normal brains incite normal people in normal circumstances to belief in God. God-beliefs are the perfectly ordinary and natural products of perfectly ordinary brain processes.

Canvassing the extreme but unsettled views we have reviewed in this chapter has not been in vain. The discussion reveals a deep and profound existential divide in looking at the brain's relationship to God: as indicative of beings or forces that transcend the brain (reality), or as nothing but the chemical, neuronal, or cognitive processes of the brain (delusion). Before we can make any progress in accepting or rejecting such claims, we need a much better sense of the science of the mind and how it is involved in production and sustenance of religious beliefs.

CHAPTER 2

Brain and Gods

Cognitive Science

From the television show *Star Trek* we heard each week: "Space, the final frontier . . ." Just as space displaced the earth as the final cosmic frontier, so, too, consciousness and the mysteries that surround it may displace space as the final frontier. While exploring the outside world with some success, we have made considerably less progress exploring inside the human mind. Cosmologists and astrophysicists gained significant ground in the past century toward understanding the nature and shape of our cosmos. We know the age of the universe, its initial conditions and fundamental constants, and how its initial configuration along with natural laws gave rise to planets and stars and even, eventually, to life. We know the extent of the visible universe and our tiny place in a corner of it in what we call the Milky Way galaxy, and we know that ours is the third rock revolving annually around the sun. We know the inner constitution of stars, the chemical furnaces and factories of the universe. And we know a lot, lot more.

Yet, after years of thought and effort, we still have no idea how the painful ache in our knees, the perception of the redness of a rose, or the sweet taste of honey arises from the brain chemicals or neural processes that are correlated with those conscious thoughts or feelings. In fact, we have no idea whatsoever how things with distinctly physical properties such as size, shape, duration, quantity, solidity, and motion (or combinations thereof) give rise to our very subjective feelings or sensations of, for example, color, happiness, or pain. How do things with entirely physical properties cause or give rise to things with entirely mental properties? What is the relationship of the mental to the physical? We simply don't know.

Here is another way to put the question: What is the relationship of subjective thoughts and feelings to the physical brain? How is my feeling of sadness the result of chemical or neural processes or events? So far, we have no idea. Brain and consciousness are the final frontier, shrouded at this point in mystery.

Yet, while we don't know how chemical or neural processes in our brains produce nonphysical thoughts or feelings, we do know, through cognitive science, a lot about how characteristic ways of human thinking incline us to fairly typical human beliefs. Cognitive science is a relatively new discipline that unites psychology, neuroscience, computer science, and philosophy into the study of the operations of the mind/brain (for purposes of this book, we will use "brain" and "mind" synonymously). Cognitive science studies how the human mind acquires, stores, organizes, retrieves, and uses information. The scientific study of the thinking mind has considered, among many other things, perception, attention, memory, pattern recognition, concept formation, consciousness, reasoning, problem-solving, language-processing, and forgetting.

It is those very cognitive faculties that have so effectively yielded understanding of the astrophysical nature of our cosmos (and which have, as yet, yielded little understanding of consciousness). Without knowing how matter produces thoughts and feelings, we know that the mind comes equipped with cognitive faculties that have allowed us to understand much of reality (including what some of those cognitive faculties are).

We have learned, as one might expect, that we have cognitive faculties that allow us to know many things. We know that there are other people (and, sometimes, what they are thinking or feeling), that there is a past and will be a future (and that the future will be like the past), that there is a world outside our minds, and that that world operates according to natural laws (best described in, say, chemistry, geology, and physics). We know arithmetic and geometry (and a special few in our species know of infinities and multiple infinities). We know that slavery is bad and that generosity is good. We have used our cognitive faculties to learn about the distant past and the unforeseen future. And we lament the millions killed in the Holocaust. Similarly, I know that there is a maple tree in my backyard, a yard filled with uncountable blades of grass, and that a robin's nest full of blue eggs is cradled in the branches of that tree.

What are some of those cognitive faculties that equip us to know such things? We have learned that we have faculties that dispose us to perceive things via our five senses and to know and judge the minds of other people (at least some of their thoughts, feelings, and desires). We cognitively assume that there was a past (we weren't created, along with the world, five minutes ago with our memories intact) and that there is an external world—both of which we can know about. We project into the future and make moral and aesthetic judgments (some of which are astonishingly universal). We are cognitively equipped and even inclined to count things; this cognitive disposition, in combination with other cognitive faculties such as those involved in reflection, led to the invention of algebra and the calculus.

We are able to think not only of the facts and just the facts, ma'am, but we are also equipped to think counterfactually—how things might be or could have been. This sort of abstract thinking allows, among many other things, scientists to think how reality might be really different from the way it appears to us. Modern science resulted when careful and creative counterfactual thinking allied with developments in mathematics. For example, while the earth appears to be unmoving, science tells us that it is moving at relatively high rates of speed in orbit around the sun. And while it looks as though the sun rises and sets, in reality the earth rotates while the sun stands still. And though it seems that tables are solid, stable, and impenetrable pieces of matter, they are really mostly empty space filled with countless atoms moving at high rates of speed. Modern science would exceed human comprehension if humans lacked the cognitive capacities to count and to think counterfactually (and, among many other things, to reflect very deeply and together over very long periods of time).

Yet our cognitive faculties mostly help us to get where we are going, avoid pitfalls, enjoy our friends and families, secure food and shelter, avoid predators and enemies, find mates, and care for our children. Our cognitive faculties equip us for astonishing intellectual accomplishments, on the one hand, and for mundane knowledge of life, people, and the world, on the other.

We also know that we are finite, fallible creatures with cognitive faculties made from the dust as it were (shaped, very likely, through evolutionary processes). We are, except for depressed individuals who may have a better grasp of reality in this regard, cognitively disposed to think we are better than average (the Lake Wobegon effect). We also typically think we are smarter and morally better than those who disagree with us, and we favor evidence that confirms our current beliefs (and we blissfully ignore evidence that opposes them). These cognitive dispositions, in turn, feed our natural dispositions to tribalism toward our in-group (people who believe and look and smell and sound like we do) and intolerance toward our out-group (people who don't look and believe and smell and sound like we do). And some human beings, using those very same cognitive faculties, believe that the earth is flat and that there are ghosts and goblins.

From cognitive science, we have learned a lot about our cognitive faculties—noble and ignoble alike—and the beliefs they produce—noble and ignoble alike. The very faculties that inform and even humanize us likewise misinform and dehumanize us.

And we have learned, in the past thirty or so years, how our minds dispose us to belief in God. Cognitive science has shown how human beings cognitively take in typical experiences and then process that information in ways that very naturally produce God-beliefs.

Gods and Brains

In addition to studying perception, attention, memory, pattern recognition, concept formation, consciousness, reasoning, problem-solving, language-processing, and forgetting, cognitive science has also studied the ways in which we acquire and sustain religious beliefs; this subdiscipline in cognitive science is called the cognitive science of religion (CSR). Every culture seems to have deeply entrenched beliefs in spiritual beings, and most have beliefs in an afterlife. And just as universal human traits such as language and emotion are explained by a mind-brain disposed to language and emotion, it is now widely accepted that universally occurring spiritual beliefs indicate that humans are naturally disposed or inclined to belief in spiritual beings. Like language, these cognitive dispositions find culturally specific expressions, but common to every culture is the firm belief in both the spiritual world and the material world.

The widespread occurrence of religious beliefs, like the widespread occurrence of perceptual beliefs ("I see a car," "It is cold") or memory beliefs ("I remember driving your Chevy on that snowy January morning"), suggests that the acquisition of religious beliefs is due to the operation of naturally occurring cognitive faculties. (I use the term "cognitive faculty" as the name of the mental tools that we use to think or cognize the world. The term was used widely by seventeenth-century philosophers who believed the major cognitive faculties were sense perception, memory, imagination, and understanding [or intellect]. We have emotive and willing faculties as well.) Just as the acquisition of language is natural, so, too, is the acquisition of religious beliefs. Religious beliefs are natural both in the sense that they are easily acquired (typically without recourse to rational reflection or argument) and because they originate with little cultural input. We are naturally disposed to religious beliefs.

What are the cognitive faculties implicated in belief in God? In the past twenty or so years, there has been a remarkable wealth of scholarship on the cognitive mechanisms involved in the production of religious belief. There is an increasingly rich science of belief in God—the cognitive science of religion. We will return again and again to this science. While CSR is in its infancy, we have achieved some understanding of the basic mechanisms involved in the production and sustenance of religious belief.[1]

The God-Faculty

Suppose you are driving on the highway and are nearly run into by a speeding, swerving car. You instantly form a belief in agency—not in the car but in a person (in a car) with the ability to act of their own accord. Non-agents, like a blade of grass in the wind or the fender of a car when hit by another car, just passively respond to forces. Agents, on the other hand, like humans and zebras and even amoebae, actively respond to various stimuli. But you do more than think of agency; you also quickly ascribe responsibility (and if you screamed, "Expletive deleted!," maybe even blame).

Now suppose the car is forced to stop at a red light and you pull up alongside. While you angrily glare at the driver you see the look on her face. Based entirely on her facial expressions, you instantly attribute fear and anxiety to the driver. What started as a mere attribution of agency and immediately shifted to an attribution of intention and purpose has now shifted again upon seeing her face. You "read" her face (correctly), forming an immediate belief about the other driver's mental state. You look around the car's interior for clues: you see a child with blood on his face and arm, and you infer that the driver is rushing her son to the nearby hospital. Your anger dissipates, replaced by sympathy.

Attributions of agency and intention equip us to *instantaneously*—non-inferentially—form the beliefs we need to respond properly in a wide variety of circumstances. We don't typically infer agency, and we don't typically reason to intention and purpose. We don't have an inner dialogue in which we argue ourselves into the belief that various sounds, sensations, or experiences are caused by agents. We don't think, "Hmm, what could that be? The car whooshed by, and most of the time when cars whoosh by me they are directed by agents. You seldom read in the news about cars operating entirely on their own. So probably that car was directed or misdirected by an agent." We simply find ourselves detecting agents. A car whooshes by, nearly clipping our bumper, and we instantly find ourselves with an agent belief.

And then we act. What moved you to action? Your agency-detecting device (ADD). When we hear a strange and unidentified sound or have a human- or animal-like visual sensation, we think of a being that can act, perhaps one that can harm us. ADD is instigated sometimes with only the slightest stimulation, immediately (that is, nonreflectively or noninferentially) producing beliefs in an agent. Being able to quickly respond to dangerous situations has obvious health benefits: sluggish responders are likely to end up dead. Our ancient ancestors who were given to reflection in such circumstances—"Let us think about this carefully: only one out of ten times does this sort of noise precede the entrance of an enemy who wishes to steal my food; thus it is vastly more likely that this is not one of those times. On the other hand . . ."—would likely remove themselves from the gene pool. Nature instead has outfitted us with cognitive faculties that produce immediate responses/beliefs independent of reflection precisely because of the urgency of these sorts of situations. Moreover, these faculties free up cognitive resources to do other things. Similar to breathing lungs and beating hearts, nature has equipped us with automatic believing and response processes that are essential to our survival.

We find ourselves almost as quickly believing (without inferring) the agent's intentions, feelings, or purposes. If it is a person, we might instantly judge that person's anger, sadness, or elation.

If it is a lion or tiger or bear, we instantly judge its intentions and then act, usually without reflection (typically we think "hostile intentions!" and RUN!!!). We automatically judge human intentions—by the look on the person's face, the shrug of their shoulders, their gait—and then we act in response. Irresponsible teenager showing off for friends or scared mother afraid for her child? Unknowing or unconcerned? Injured or arrogant? Studies show that human beings are, overall, pretty good at reading people's intentions from their faces. "Mindreading," Simon Baron-Cohen calls it.[2]

Let us change the example. Consider your response when you are home alone, in deep sleep, and are awakened by a sharp sound downstairs. You are immediately ready to spring into action. Your adrenaline has caused your heart to race. Your muscles have tensed. You act—you bolt upright, quickly reach under your bed for your baseball bat, and holler out, voice slightly atremble, "Who's down there?" You have acted instinctively with little conscious reflection in response to the sound—you believe someone is down there. You heard a strange noise and believed immediately, without thinking, that there was something, someone even, down there who might do you harm. While living in a relatively safe culture has desensitized our ADD and dulled our reactions, being instantly frightened by a strange noise in the house reminds us that our ADD has simply been lying dormant. It takes just the slightest nudge to awaken our slumbering ADD.

Of course, a sensitive ADD will produce many true beliefs that elicit appropriate fight-or-flight responses; but it will also will produce some *false positives*. ADD sometimes cries "Wolf!" Although ADD is a pretty reliable belief/action cognitive faculty, because of its hair-trigger sensitivity, ADD has also been called the "hypersensitive agency-detecting device" (HADD).

But awaking our slumbering ADD is not sufficient to move us to act. Do we stand there and hold our ground or flee to fight another day? Is it enemy, predator, or friend? In addition to ADD, we are equipped, as noted, with the ability to grasp another person's inner or mental life based on what we see of their face and body. This cognitive faculty has various names, from theory of mind (ToM) to mindreading. ToM does not settle for the postulation of mindless agency; ToM searches out the agent's intentions or purposes. Only if we are able to effectively postulate the agent's intention(s) can we rationally act. Only by judging if they are angry or hungry or friendly can we act accordingly. Returning to the example above, as soon as you started ascribing beliefs and desires to the agent in question, you kicked cognition into another gear called theory of mind. Your agency-detecting device ascribed agency to the sound, and then theory of mind took over as you tried to understand the reason the agent was acting: Is the agent looking for shelter or looking for food? Again, while at times ToM ascribes purpose to things that couldn't have intentions (to faces in the clouds, the distribution of stars in the sky [astrology], or the plumbing noise that woke you up), it is an enormously useful cognitive faculty for instantly ascertaining purpose (which, in turn, allows us to quickly and effectively act in response).

In the normal course of human life, we perceive human forms, faces, and voices, and ADD automatically, nonreflectively attributes human agency as the source. "That is an agent," we think. Then a second cognitive system responsible for generating inferences related to mental states—the theory of mind—is activated and automatically fills in details about the (human) agent's likely beliefs, desires, emotions, and so forth. That agent has this intention. By simply looking at another person's face, we are pretty good at judging their anger, fear, embarrassment, or sadness. From the intonation in another's voice, we can detect irony and sarcasm. By watching a person

25

walk, we can "see" that they are joyously buoyant or weighed-down-by-the-world depressed.

ToM, like ADD, is not infallible, of course. We make mistakes; people lie and hide their real feelings. Nonetheless, most normally functioning human beings are pretty good at "mindreading"—instantly judging another person's thoughts, feelings, and desires.

We also easily attribute intentions when they aren't there.

One snowy day, upon my return from the grocery store, I slipped as I exited my car. As I fell, the bag of groceries I was holding spilled out onto the icy ground. At that moment, my wife opened the back door of our house, and our dog, Wrinkles, bounded out to greet me. For a brief moment I had this Lassie-like thought: "He sees my distress and is rushing to my rescue. He'll nuzzle me up off the ground and to safety." As Wrinkles got closer, though, he ran right past me and started eating the hamburger meat that had fallen out of my bag. My ToM had attributed human-like intentions to an animal that had, if any at all, totally different intentions. We have a powerful urge to anthropomorphize—to ascribe human intentions to nonhuman objects and events—an urge to which I unwittingly acceded with respect to my dog.

It is easy to misattribute agency and misapply intention. The agency-detecting device and theory of mind, fine as they are, are anything but infallible. Imagine walking through the woods at dusk and hearing a sharp noise off to your left. You quickly turn and see, off in the distance, a face staring menacingly at you from afar. Your fears become palpable as your adrenaline pumps and your heart rate increases. You wonder about that person's intentions (wayside robber or fellow traveler?). Your sweating brow indicates that you are erring on the side of caution—robber, not friend. You shine your flashlight directly on the face and see it for what it really is: a gnarly knot in a tree. Your fears slowly subside, but it will be a while before your blood pressure returns to normal. ADD and ToM, in this case, got it wrong.

We see agents when they aren't there, and we attribute intentions when they aren't present (sometimes to things or events that don't have intentions). We hear a rustling bush and think *lion*. We see a lion and think *courageous*. We see a friend approaching but think *enemy*. Our crops get flooded and we think we are being punished. And so on and so on. ADD and ToM are reliable but not infallible. It is precisely this misattribution of agency and intention that some cognitive scientists think produces belief in gods.

How, then, did human beings get from agents and intentions all the way up to gods?

Some cognitive scientists speculate that ADD (of the hypersensitive variety—HADD) and ToM combined, through a series of misfirings, to produce beliefs in gods. Occasionally, sounds, shapes, patterns, or movement trigger a "HADD-experience," an attribution of agency that is inconsistent with any known natural agents such as humans or animals. In these situations, the activity of HADD may lead to the postulation of a different sort of agent such as a god. Psychologist Justin Barrett explains:

> Our minds have numerous pattern detectors that organize visual information into meaningful units. HADD remains on the lookout for patterns known to be caused by agents. If this patterned information matches patterns . . . known to be caused by agents, HADD detects agency and alerts other mental tools. . . . More interesting is when a pattern is detected that appears to be purposeful or goal directed and, secondarily, does not appear to be caused by ordinary mechanical or biological causes. Such patterns may prompt HADD to attribute the traces to agency yet to be identified: unknown persons, animals, or space aliens, ghosts, or gods.[3]

According to this view, misattributions of agency and misapplications of intention or purpose generate beliefs in gods.

In discussions of the cognitive origins of religious belief, ADD and ToM, taken together, are often called the God-faculty.[4] This God-faculty, with little or no conscious reflection, moves us to God-beliefs and religious practice.

Though tailored by natural selection for a particular domain of activity, the God-faculty's flexibility and hair-trigger tuning makes it liable to produce beliefs in unseen agents or intentional agents with other supernatural properties.[5] Rather than deductively reasoning to the existence of an intelligent being to account for mysterious bumps in the night or faces in the clouds, anthropologist Stewart Guthrie argues that human cognitive systems are tuned to rapidly intuit the presence of intentional agents in the environment, even given scant or incomplete evidence.[6] Under certain conditions (those in which we are tempted to make sense of—to humanize—our ambiguous and confusing natural world), this tendency may generate beliefs in anthropomorphic gods, says Guthrie.

What specific properties might these gods be prone to have? Much like humans, gods will have percepts, thoughts, beliefs, desires, goals, motivations, and emotions. Likely, they will have language, communication, and social relations. All of these basic attributes come automatically from ToM. Though Guthrie emphasizes the attribution of human properties (hence, he regards his ideas as a new anthropomorphism theory of religion), he admits that the cognitive faculties at play invite more flexible input conditions than distinctly human agency. After all, they must accommodate nonhuman animals (such as saber-toothed tigers) and disguised or camouflaged agents. Since HADD is able to handle nonvisible agents, the idea of an *in*visible god does not deter the God-faculty at all.

Sometimes humans have postulated extraordinary agents with extraordinary powers that act for extraordinary reasons through apparent miracles, floods, or thunder. Such big and powerful agents have big reasons for the things they do. If one has acquired beliefs in extrahuman agents with super-qualities—super-powers and super-knowledge, for example—one has a ready explanation for the causes of super-events. In short, ADD and ToM have produced beliefs in minded supernatural agents that have had rich potential for explaining some very important life issues (the weather, for example, or success in war).

"God-faculty," then, is unduly restrictive. By "gods," I mean any supernatural intentional agents or powers, often disembodied spirits, whose existence would impinge upon human activity (could be Allah, Baal, ghosts, goblins, or fairies).

Moreover, the God-faculty is not dedicated to producing God-beliefs. It is, if anything, dedicated to getting us to act appropriately (fight, say, or flee) in response to an agent with intentions. Insofar as it produces beliefs instead of actions, it is "aimed at" producing very ordinary beliefs in very ordinary agents such as lions and people. But, in addition to its typical lion- and person-belief outputs, this collection of human cognitive faculties also non-reflectively produces beliefs in gods (given ordinary inputs from ordinary environments). CSR holds that we do, indeed, have a natural, instinctive religious sense and that the beliefs it produces are typically noninferential, ordinary, and natural. Belief in God is the perfectly normal product of our perfectly normal and naturally functioning cognitive faculties.[7]

HADD+ToM are activated by specific experiences, but God-beliefs aren't preexisting tacit assumptions waiting to be activated; God-beliefs are constructed in response to particular environmental stimulations. These experiences could be bumps in the night, faces in the clouds, or striking cases of fortune or misfortune. The God-belief that gets triggered will have human-like mental attributes but may also have supernatural powers, may be invisible, and may perhaps be morally interested in human affairs. Exactly which properties the god possesses—cosmic creator, super-knowing, super-perceiving, immortal, immutable, or wholly good—are largely unspecified.

While ADD and ToM are usually the key players in the CSR story of the production of religious beliefs, other cognitive faculties are likewise implicated. While not denying the role of HADD-experiences in generating or encouraging belief in gods, psychologist Jesse Bering has begun developing a variant on the God-faculty that allows for a broader range of experiences to trigger or excite thoughts about gods. On the basis of some experimental evidence with children, Bering argues for a pan-human, early-developing tendency to wonder "What does it mean?" or "Why me?," especially with regard to striking experiences of fortune or misfortune. Bering argues that when faced by these sorts of experiences, we automatically speculate about the *intentions* of an unspecified agency that might account for the event and what the event means. Bering has dubbed this cognitive tendency the existential theory of mind, or eToM.[8] In addition to HADD-experiences, events of striking fortune or misfortune, strange coincidences, and the like prompt us to consider the intentions of the some*one* who has orchestrated the event. In this way, eToM, like HADD, may stimulate thinking about and belief in gods—gods that influence human affairs, perhaps rewarding or punishing in a morally concerned way. Indeed, Bering and colleagues have suggested that the tendency to attribute events to the activities of morally concerned intentional agents, and hence belief that gods are watching as potential moral police, may be an adaptation that assisted in building cooperative communities.[9]

Another pathway to belief in gods may hinge on representations of death. At my father's funeral (he passed away very suddenly and at a young age), I saw his lifeless body but also found myself thinking that he would just wake up and start talking to me again. When I returned to my childhood home, I stared at his empty chair and found myself thinking that he would soon be sitting in it, falling asleep with the TV on as he did most every night. In late November, I walked in his garden, wandering through the decaying tomatoes and rotting broccoli, and "spoke" to him in my mind. I later learned that I was not alone in postmortem conversations with the deceased.

Bering argues that we intuitively continue to attribute mental states to minds or spirits that we believe survive death.[10] For example, after people die, many of us continue talking with our loved ones in our minds as though they were still existing. Though they are not present in body, we continue to commune and communicate with those who are "present" in spirit. This difficulty of mentally simulating the cessation of mental states makes the idea of minds or spirits surviving death intuitive.[11] It is not surprising, then, that one of the most widespread and perhaps oldest kinds of God-belief is in ancestor spirits and ghosts.

Bering's claims are supported by studies that contend we are natural mind-body dualists: we naturally believe that humans are a composite of a physical body and a nonphysical or spiritual soul. Psychologist Paul Bloom, among many others, argues that we intuitively understand persons as composites of both physical bodies and nonphysical souls and that this dualism about persons undergirds the rather natural belief in the survival of one's spirit after the death of one's body.[12] Believing that disembodied minds of the deceased continue to exist and interact with humans, then, is a nonreflective product of ordinary cognitive systems.[13] And, again, it is a thin conceptual line that separates belief in ancestor spirits from, say, belief in Yahweh.

Convergent with these findings, developmental psychologist Deborah Kelemen has suggested that children may be "intuitive theists," on the basis of a series of studies regarding children's maturationally natural cognition relevant to understanding the causes of things in the natural world.[14] In brief, research suggests that children have what Kelemen calls "promiscuous teleology," favoring design- and purpose-based accounts of natural phenomena, beyond what they might have been taught. Hence, four-year-olds are happier with teleofunctional accounts of why rocks are pointy (e.g., so that animals will not sit on them) than with mechanistic accounts (e.g., because bits of matter piled up over time and wind and rain shaped them). Further, they assume that intentional agents, not mechanistic causes, bring about design and order. A tendency to see the natural world as designed, together with an intuition that design entails or assumes intentional agency (design implies a designer), leads children to readily embrace creationism and other types of supernaturalisms with regard to understanding the natural world (and to reject evolution and materialism).

Additional research on children's understanding of minds suggests that their default settings as preschoolers is that others' minds are super-knowing, super-perceiving, and (perhaps) immortal.[15] Children naturally assume superhuman capabilities on these dimensions when applied to people, to many animals, or to gods. They then pare back the super-ness as they learn about human perceptual fallibility, limitations on knowledge, and mortality. In this respect, a god that has super-knowledge, has super-perception, and/or is immortal requires less learning for preschoolers than learning about humans. Due to the default settings on their conceptual systems, children are "prepared" to understand and affirm many aspects of a super creator God.[16]

A Summing Up

The agency-detecting device (ADD) and theory of mind (ToM) combine to create widespread beliefs in godlike agents (faces in the clouds). But cognitive faculties shape other religious beliefs as well: mind-body dualism, an afterlife of actively involved spirits, and promiscuous teleology. These sorts of findings from developmental psychology coalesce to lead scholars to agree with Bloom that "religion is natural."[17] Leading thinkers in cognitive science of religion concur. Psychologist Justin Barrett argues, "Belief in God or gods is not some artificial intrusion into the natural state of human affairs. Rather, belief in gods generally and God particularly arises through the natural, ordinary operation of human minds in natural, ordinary environments."[18] Pascal Boyer, as noted, contends that religious beliefs are natural: religious beliefs are naturally formed by faculties of the human mind.

Cognitive science, it should be noted, favors no particular religion or set of God-beliefs; it is not especially supportive of, say, Christianity or even of the Abrahamic religions more broadly. Just as we are disposed to acquire a language (but not, say, English), so, too, we are disposed to acquire religious beliefs and practices (but not, say, Christianity). There is no genetic or cognitive determinism here: having a natural disposition to acquire a language leaves a lot to culture to influence *which* language (language is not all in our genes), and having a natural disposition to acquire religious beliefs/practices leaves a lot to culture to influence which religious beliefs/practices (religion is not all in our genes). And just as our cognitive faculties don't prefer or favor or valorize any particular language, they don't prefer or favor or valorize any particular religion (or religion at all).

Cognitive science holds that our minds come equipped with cognitive faculties that actively process our perceptions and shape our conceptions of the world.[19] These common cognitive faculties structure, inform, enhance, and limit the way we think about the world around us. Some of these cognitive faculties structure, inform, enhance, and limit religious beliefs. Belief in gods, to take one characteristic form of religious belief, arises from the stimulation of our cognitive faculties. That is, because of the characteristic architecture of human minds, independent of special enculturation, humans are receptive to the existence of gods and readily reason about their activities and form collective actions (such as rituals) in response to these beliefs. From the perspective of cognitive science of religion, beliefs in gods are natural products of our common cognitive faculties, and in this sense religious belief is "natural."

Evolutionary Explanations of Religious Belief

That humans have a God-faculty (ADD and ToM) seems undeniable. Some have gone further, attempting to discern how the God-faculty, like many of our other cognitive faculties, was acquired evolutionarily. Explanations of the God-faculty often appeal to evolutionary explanations of the origins of our cognitive faculties. I have already hinted at these explanations in the previous section, but they bear repeating. *Evolutionary explanations* of our cognitive faculties seek to account for the development of our various mental faculties and concepts in terms of responses to the various pushes and pulls that our primitive ancestors experienced. Just as the opposable thumb was shaped by a process of evolution through natural selection, our brains (minds) developed similarly and, when successful, increased our reproductive success. Put simply, evolutionary psychology holds that the human mind developed in our primitive ancestors through natural selection to solve adaptive problems in their ancestral environment.

Our brains/minds developed, in response to various selective pressures, a set of intellectual tools that fit us for survival in a world with enemies and friends, kin and clan, natural disasters and regular seasons, and predators and prey. Our cognitive faculties were cobbled together, as responses to various adaptive problems, in ways that enabled us to perceive middle-sized objects (not the astronomically large or microscopically small) like people and pigeons; to take into account past, present, and future; to anticipate plans of our enemies and possible mates/friends; to contemplate, initiate, and complete plans of our own; and to communicate, feel, and remember. These and other cognitive abilities developed in response to environmental pressures in our hunter-gatherer past in accord with our most fundamental human needs (the famous four Fs—feeding, fleeing, fighting, and reproducing).

While we once thought animals were ruled by instinct and humans by reason, we now understand that our evolutionary history has shaped us so that a great deal of human cognition and action are instinctual (immediate, noninferential, nonreflective). We are superior to the animals, perhaps, not because we lack such instincts but because we have so many more of them. We have acquired, evolutionarily, countless (mostly invisible) instincts: the ways we see and what we hear, what we find attractive or repulsive in smells or food, who or what to fear and when to flee or fight, what to say and to whom and when (which assumes our instinctual ability to acquire languages), what is beautiful (or ugly), and what is good (or bad)—among many, many other cognitive and practical instincts. (This is not to deny the huge role that culture plays in the specific content of beliefs and the shaping of practices.) Just as we don't decide to breathe or make our hearts beat, neither do we decide, in the vast majority of cases, what to believe or what to do. Given various informational inputs, our cognitive faculties immediately, noninferentially produce beliefs and/or actions.

Not surprisingly, then, humans have, for example, a deep-seated and instinctive fear of snakes; no doubt our ancestors who had such fears lived longer (and thus passed on their snake-fearing genes to succeeding generations) than those who did not. Humans have natural aversions to, among other things, incest and feces. While we might think the former a distinctly moral judgment, inbreeding typically lowers physical and mental prowess (thus decreasing reproductive success), and thus our natural aversion to incest likely has evolutionary roots. And our ancestors to whom feces stank (who thus distanced themselves from human waste) avoided more waste-borne diseases than their less discriminating brothers and sisters (again, increasing reproductive success).

One might expect, then, to find a coterie of cognitive faculties directly related to mating, for example, and avoiding enemies. And we do.

Consider the psychology of mate preferences. Women desire men high up in the social hierarchy (such men are more likely to be able to care for offspring), so men competitively assert themselves to achieve cultural success (thus explaining, in one fell swoop, the attraction of beautiful young women to rich old men). Culturally and financially successful men are more likely than poor and dispossessed men to have and successfully raise children (and so pass on their competitive genes). Judgments of physical attractiveness likely served as barometers of the health of a possible mate; those with healthier partners are more likely to produce and raise healthy offspring, thus passing on healthy genes to future generations. Men have a tendency to overestimate a woman's interest in having sex with them (perhaps giving them confidence to approach potential mates), whereas women have a tendency to overestimate a man's interest in sticking around after the baby is born. Since the evolutionary costs of pregnancy and raising a child are disproportionately high for women, men are assumed to desire casual, uncommitted sex more than women.

What sorts of cognitive mechanisms did we develop with respect to those other humans who are in competition with us for scarce resources—that is, (possible) enemies? Life on the Serengeti was kinnish or clannish, with fierce competition between kin or clan groups. The ability to quickly judge friend or foe could mean the difference between life and death. It seems, then, that we have a natural fear of strangers and even more so of male strangers. This fear manifests itself when one is walking alone down a dark street and sees a large stranger slowly walking toward one. The "stranger" the person (larger and dressed differently, say, than those in one's own community), the more likely we are to judge that they will harm us. If we are able to see their face, we are highly skilled at judging anger or friendliness, thus quickly assessing threat and our possible responses.

Likewise, we have acquired agency detection and theory of mind instincts. We have already discussed the evolutionary advantages of ADD to our ancestors. The ability to detect agents quickly and without inference, even given the possibility of false positives, was eminently useful. ToM very likely developed in our primitive predecessors who needed to better negotiate tricky relationships with human competitors. The better we are at detecting the desires of our human competitors, the better we can plan and act accordingly. Are they hostile or friendly? Are they hungry or satisfied? Are they responding with fear or anticipation to our advances? Does that possible mate find me attractive or repulsive? Without some ability to make speedy judgments about the intentions of other people, humans would not respond instantly and plan accordingly. ToM instinctively produces in us beliefs about the purposes of minded agents. There are similar stories one could tell about the adaptive advantages of, for example, teleological and existential reasoning.

According to some, religious beliefs are an evolutionary problem. With respect to adaptive fitness, many religious practices seem *maladaptive*; in evolutionary terms, they don't seem very conducive to reproductive success. Atran, when discussing the costliness of religion, calls it "materially expensive" and "unrelentingly counterfactual and even counterintuitive." "Religious practice," he continues, "is costly in terms of material sacrifice (at least one's prayer time), emotional expenditure (inciting fears and hopes), and cognitive effort (maintaining both factual and counterintuitive networks of beliefs)."[20] For example, severe religious practices, such as celibacy and the sacrifice of virgins, run counter, to say the least, to reproductive success. A religious group that consistently practiced both would be quickly removed from the gene pool. The Shakers, who forbid sexual activity, commit evolutionary suicide. Likewise, in times of scarcity (which were most times for our primitive ancestors), sacrificing highly valuable commodities such as grain and animals is hardly conducive to survival. Even practices like worship and prayer can be costly, because they take time away from hunting, gathering, and reproducing.

So how could such costly practices catch on? How is it that, throughout human existence, belief in spiritual reality has been the norm? How could behaviors so costly survive the precise but cruel culling of natural selection? There are two accounts of the evolutionary origin of religion: one holds that religion is a by-product while the other holds that religion is adaptive.

Sometimes natural selection produces a trait that is a *by-product*, not a direct consequence of natural selection. That is, natural selection selects for an adaptive trait, a trait that increases one's reproductive success, but the trait is accompanied by another trait that is nonadaptive (not maladaptive, simply not adaptive); on its own, the by-product would not have been selected. In humans, the redness of blood is a by-product of hemoglobin's ability to store oxygen (hemoglobin turns red when oxygenated). And the wrinkles on our knuckles are by-products of our evolutionarily successful ability to bend our fingers. By-products are accidental, nonadaptive leftovers; they aren't adaptive traits.

A by-product belief, to coin a term, is a belief that is a by-product of faculties "designed" for the production of other sorts of beliefs. If the standard picture is correct, religious belief is a nonadaptive by-product belief.

Religious beliefs arise, then, as by-product beliefs of our otherwise well-intended and survival-conducive agency-detecting device. Scary, big, and/or portentous phenomena that cannot be explained by either human or beast are attributed to spiritual (that is, nonphysical) agents with super-powers and, through ToM, various intentions. These nonphysical beings, then, are either for us or against us, and we need to begin devising plans to appease or please them (thus the origin of religious ritual). ADD creates belief in nonphysical agents, and ToM embellishes that initial belief into (anthropomorphic) gods.

Our predator and enemy detector went considerably further afield to produce belief in gods, turning to gods to "explain" the weather, motions of the planets, success in hunting or growing crops, good and bad fortune, disease, and even death itself.[21]

At other times natural selection produces a trait for the purpose of *adaptation*. Is it possible that religious beliefs are adaptations that do indeed contribute to fitness? Although religious belief is typically conceived of as a by-product belief, there are important dissenters and good reasons to dissent. Joseph Bulbulia, for example, claims that "we need to begin thinking about our religious traditions not as mistakes and costly maladaptations but as practices for human flourishing."[22] How, then, might religious beliefs prove adaptive?

Because religious beliefs often engage passions and emotions, they can be highly inspirational and motivational. One's ADD and ToM might function repeatedly to produce lots of person-beliefs when one is walking in the mall. But these are fleeting and casual, and they seldom place any moral demands on us. Religious beliefs, on the other hand, create a sense of obligation or unworthiness toward divine persons. And they engage us deeply because they provide compelling reasons, reasons that engage our passions and desires, to be moral. It is well known that religions often supply the content of morality. But they also *move* us to be moral. Religious beliefs give (and gave) morality its oomph.

In *Big Gods: How Religion Transformed Cooperation and Conflict*, Ara Norenzayan argues that "Big Gods" is one solution to the puzzle of early human cooperation.[23] Big Gods, which came to dominate the human cultural landscape, bound early humans together into increasingly large societies by securing the cooperation of those within and increasing success against those without (that is, against competing groups that lacked Big God concepts). It is not hard to explain on evolutionary grounds exclusive and selfish concern for one's self or even cooperation among kin; kin, after all, share one's genes, and evolution is all about getting one's genes into the next generation. And it is not hard to explain a quaking fear of non-kin in competition for scarce resources. It is hard to explain on evolutionary grounds how early humans were able to overcome their quaking fear of non-kin and to begin behaving cooperatively in increasingly large (and successful) human communities. How deeply selfish and competitive individuals cooperated with other deeply selfish and competitive unrelated individuals in large communities is, in a nutshell, the evolutionary puzzle of human cooperation.

There are, to be sure, cooperative benefits to be had from living in large human communities. More game can be caught by groups of hunters than by individual hunters. Collective action against one's enemies is vastly preferable to fighting alone. Shared labor and specialized expertise ensure a bountiful harvest in agricultural communities. Having and caring for a child on one's own is extremely costly in evolutionary terms, so sharing childrearing is a tremendous benefit, especially to women. There are more possible mates within larger communities. And better food at potlucks. In short, it pays to cooperate, but (and here the puzzle of human cooperation rears its ugly head once again) *only if you can trust the other selfish and competitive members of your community* (and they can trust you). How, then, can one ensure the trust necessary for selfish and competitive individuals to live together in peace and to gain, without threat of loss, those cooperative benefits?

Here is where religion seems to have helped. How might religious beliefs have motivated early humans to cooperate? God-beliefs can motivate cooperation if one believes that God sees everything that one does (so one can never get away with doing evil). One could surely satisfy one's primitive desires better (get more food, have more sex), even within a large, otherwise cooperative community, if one could steal and rape and get away with it. If one believes that there are only limited, human persons to fool, one might reasonably calculate that stealing food or killing such persons is, on occasion, in one's best interest.

Suppose you are a slave on a Greek galley ship, one rower among forty. You cultivate the ability to sweat profusely, earnestly wrinkle your brow, and flex your sinewy muscles—all the while minimizing your output. The other thirty-nine, after all, are pulling their weight. Because you work less than your fellow rowers, you burn fewer calories and require less sleep. You eat your allotted gruel, the exact same amount as the rest. When they, fully expended and exhausted, fall into a deep sleep, you sneak some additional food. Because you work less and eat more, you are considerably healthier. Some of your overworked and underfed partners are too sick to mate; some just plain die young. But you, in full vigor, enjoy the port city nightlife, spreading your seed widely. Your selfish genes are passed on, while their cooperative genes are removed from the gene pool.

Of course, every time the galley master stares at you, whip at the ready, you work just as hard as everyone else. And when the food stocks are monitored throughout the night, you miss your evening snack. But when you aren't being watched, you freeload on your more cooperative compatriots.

It is easy to see, from this example, the evolutionary puzzle of cooperation. Selfish freeloaders, giving less and taking more than their fair share from a group of cooperators, are more likely to pass their genes on to succeeding generations. And so we should expect, evolutionarily, dispositions to selfishness to trump dispositions to cooperation in the long run.

But maybe God is like the galley master.

If you are part of a community that affirms an all-knowing, perfectly just God, you will think that you can never escape detection: unlike limited humans, God *always* knows what you are doing. If you and your community members believe in God, it is never rational to think that you can be immoral and get away with it. With God always looking over your shoulder, it is a good idea to abide by the rules. In short, God-beliefs can motivate altruistic behavior.

Interestingly, cognitive science has shown a natural human tendency to regard supernatural agents as having super-knowing and super-perceiving powers. A merely super-knowing God is not sufficient to prevent people from being immoral. People must also think that the super-knower also has the powers to punish them. Perhaps the gods mete out their rewards and punishments in this life through fortune or misfortune. The gods are often conceived of as punishing people here and now through a stubbed toe, a plague of locusts, a flood, or a disfiguring accident. Or the punishment may be conceived of as being meted out in the next life, where the scales of justice are balanced. Interestingly, cognitive science has shown that we have a natural tendency to attribute purposiveness to fortune and misfortune and a natural tendency to believe in a next life.

Cognitive scientists "peek" into the workings of the human mind through various priming techniques. When experimenters subtly instill concepts (called "priming"), they seek to elicit various unconscious responses. In psychological experiments, priming occurs below the conscious level of thought. By exposing the subject to one stimulus, experimenters influence responses to another stimulus. For example, subjects primed (unconsciously stimulated) by assembling puzzles that contain a lot of the color purple might "see" grapes when shown (conscious stimulus) pictures of circles. Subjects who are primed by unscrambling word puzzles containing terms like "flag" and "country" might elicit feelings of patriotism when later stimulated by videos of their fellow citizens winning in the Olympics. Subjects who are told that a bottle of wine is expensive (the prime) are vastly more likely to think the wine is very good when tasting it (the stimulus) (and to think the wine is bad when told that it is cheap). A host of empirical studies, many conducted by Norenzayan and colleagues, suggest that people are kinder, cheat less, share more, and are more cooperative when they believe they are being watched. The mere presence of a pair of eyes on a computer screen decreases cheating on tests; in the same way, the mere presence of a pair of eyes on a tin can increases paying for sodas in an honor system (placing a dollar in the tin can outside the refrigerator).

And those are just eyeballs. What about God primes? When experimenters subtly instilled supernatural concepts in the minds of college students, college students cheated less on tests: those to whom the experimenter nonchalantly mentioned that the ghost of a dead graduate student had been seen in the testing room cheated less than those to whom the ghost story was not mentioned. When primed with God-concepts, people share significantly more money with anonymous strangers (without hope of any return). Individuals in both the control and the primed groups were given $10 to share as they pleased with an anonymous stranger (whom they never saw). Before divvying up the money, individuals in both groups unscrambled sentences; the primed group's sentences contained the words "spirit," "divine," "God," "sacred," and "prophet," while the control group's sentences had neutral terms. The God-primed group offered an average of $4.56 (virtually perfect other-regard), while those in the neutral condition offered $2.56. Even atheists were more generous when they were religiously primed. When we believe God is watching us, we are decidedly more cooperative.

CSR has likewise shown that fear of punishment, especially of the divine variety, effectively motivates cooperative behavior. While benign or nice spiritual beliefs can and often do motivate cooperative behavior, mean God beliefs are even more motivating. Belief in a punishing supernatural agent reduces (and reduced, historically) social transgressions more than belief in a loving supernatural agent. As Norenzayan puts it, "Hell is stronger than Heaven."[24] Dominic Johnson and Oliver Krüger, who call this "supernatural punishment theory," argue that divine sticks increase cooperation significantly more than divine carrots.[25] In a series of studies, Norenzayan found that, while belief in a compassionate God did not effectively reduce cheating, belief in a punitive God did (and did so remarkably well). In short, the belief that God both monitors human behavior and punishes the wicked seems highly effective in reducing social transgressions and encouraging cooperation. Supernatural punishment theory has been vindicated not just in experiments with first-year college students but sociohistorically as well. In an exhaustive survey of 186 societies around the globe, Dominic Johnson found a significant correlation between moralizing high gods, on the one hand, and compliance with social norms, on the other.

Supernatural punishment theory makes good evolutionary sense. While there are dramatic benefits to cooperation, those who cheat (steal food, don't share, are lazy, etc.) and get away with their transgressions are threats to the livelihood of the community. Assigning humans to police one's community and to punish transgressors is costly: the duties of policing and punishing prevent otherwise capable people from fighting, fleeing, feeding, and reproducing. However, since belief in a moralizing high god effectively increases cooperation and reduces transgressions, the cost of policing and punishing is dramatically reduced. An all-knowing god sniffs out transgressions wherever and whenever they occur and punishes transgressors for free.

Religious beliefs, however, are seldom cost-free. Even the simplest religious rituals, such as prayer and worship, take time away from the four Fs. Sacrifices of grain and game to gods run clean counter to the satisfaction of our most basic human needs. Philosophers, of course, are obsessed with beliefs and unconcerned with rituals. But since religion involves practices as much as beliefs, any thorough or even adequate discussion of the evolution of religion must account for religious rituals. Since I am primarily concerned with the rationality of religious belief, I will briefly discuss just one theory about ritual and the origin of religious belief—costly signaling—which is intimately connected to religion and cooperation.

How can we tell who is a member of our community and who is trustworthy? How can we know who is a sincere cooperator? Knowing whom to trust is exacerbated by the ease with which some people can fake sincerity. Surely the best option for fulfilling one's own evolutionary needs is to deceptively "cooperate" while freeloading as best one can (without getting caught)—to work as little as one can (yet appear to be hard-working) and to take as much as one is able (yet appear to be taking one's fair share); all the while, other members of one's group are working their hardest and taking just what they are owed. If we can be sure that people are sincere cooperators (at least be sure that they aren't freeloaders), we can be fully motivated to work hard for the common good and sleep well at night to boot. Are there any social mechanisms that can discourage freeloaders and encourage genuinely cooperative behavior? Costly signaling theory holds that engaging in various costly or sacrificial actions demonstrates to others one's moral sincerity, thereby encouraging them to reciprocate.

Some signals of commitment, often religious rituals and taboos, are so severe and costly that one would be foolish to make them without being a sincere cooperator. For example, some religious initiation rituals, such as those involving scarification or tattooing, are so painful and even dangerous that one would not reasonably undergo them unless one genuinely wished to be a fully contributing member of that community. Such one-off rituals are major markers of sincerity. Less severe but more regular rituals—daily prayers, tithing, avoiding certain kinds of foods, etc.—likewise signal one's willingness to submit to the good of the community, thus promoting cooperation, commitment, and mutual trust. Richard Sosis's study of nineteenth-century utopian communities demonstrated that religious communes with the most costly requirements of their members lasted considerably longer than either secular communes or communes with less costly requirements. Costly religious rituals, unlike secular rituals, promoted intragroup cooperation and cohesion.[26]

A much larger set of cognitive faculties is involved in creating and sustaining religious beliefs than we initially imagined—ADD, ToM, emotion, motivation, belief in a next life, belief in purpose, and morality. Let us call this set of faculties *the mega-God-faculty*, because the beliefs that it produces are about a million times bigger and stronger than the beliefs produced by the basic God-faculty (ADD plus ToM). Moreover, the mega-God-faculty often operates within cultural contexts of religious rituals and taboos, which increase the motivational oomph.

Since cooperation proved essential to the evolutionary success of *Homo sapiens*, we need to consider those human cognitive capacities that equipped us to live together in harmony. The mega-God-faculty, when allied with various religious rituals, seems very likely to have served that purpose. A moral sense supplies information about what one should do in order to live harmoniously in community (don't lie, cheat, kill, or steal, for example), and mega-god beliefs passionately motivate self-interested people to be moral. Communities of cooperators are more likely than disharmonious communities to contain individuals who live longer and better. So the mega-God-faculty and the religious beliefs that it produces are likely to be adaptive and not by-products. Religious beliefs, though apparently costly, are conducive to human flourishing. The survival advantages of a community that has the full explanatory resources for the reward of good behavior and the punishment of bad behavior to ensure cooperation and thereby produce cooperative benefits are obvious.

Yet

We can be pretty sure that, since most religious beliefs of most people throughout most of human history involve beliefs in personal agents, ADD and ToM are involved in the production and sustenance of religious belief. Just as we use several of our senses—sight, say, and touch—to cognize trees (their variegated colors and size, the rough feel of their bark and the smooth feel of their leaves), so, too, we cognize gods with, among many other cognitive faculties, ADD and ToM. Of course, God-beliefs are considerably richer than mere attributions of supernatural agency with minimal intentions. We ascribe purposes to gods and universes, combining them in some cases into the grand, divinely instilled purpose of it all. We care deeply about our place in the lonely cosmos and think about the meaning of life—our own divinely instilled purpose. Since most of our gods are big and our religious rituals costly, God informs morality, polices humanity, threatens punishment, and motivates cooperation.

God-beliefs are likewise inference-rich. When our reasoning faculty takes them on, God-beliefs can be extended to the whole of reality. Our natural causal intuitions may lead us to wonder about the First Cause of the universe or Prime Mover of the stars and planets. Our natural inclination to see things in terms of purposes may lead us to deeper beliefs in a Divine Purposer. Our deep moral commitments may, upon reflection, suggest an underlying Moral Reality that transcends the physical world. In short, our other cognitive faculties and our deepest intuitions may either directly produce more highly sophisticated religious beliefs or take the shape of more formal arguments for the existence and nature of God.

Evolutionarily, all of the above must be understood within the complex context of reproductive success and human culture. And we have said very little about the influence of culture and of the coevolution of humanity and culture. Let us suppose that the modern human brain was capable of sustaining religious cognition starting 100,000 years ago. What exactly happened with respect to religious belief and practice from, say, 100,000 BCE to the time of the Hebrew prophets, Socrates, and Confucius is, to say the least, unclear. The evidence is scanty to nonexistent. We don't know what those earliest religious beliefs and practices were, how they were formed, and by which cognitive faculties. We don't know if they were adaptive or by-products. Evolutionary explanations of religion are both fascinating and frustrating. We really know very little about the evolution of religious belief.

The cognitive science of religion is in its infancy, and the evolutionary psychology of religion is even younger. Both are exciting and challenging fields. At this point we can safely say, I think, that we have some sense of the cognitive faculties that are involved in the production and sustenance of religious belief. But, since science is science, it is always partial and preliminary. Some of the CSR research presented will be rejected, some will be revised, and some will likely endure.

Some reject evolutionary psychology completely. They claim that since brain tissue seldom fossilizes, there is precious little hard data on which to build such theories. Theory-building in evolutionary psychology is, one might think, a series of "Just-So Stories": as with the elephant's nose, we know what needs to be explained—say, belief in God—so we invent an evolutionary story to explain belief in God. Since there is no hard data, an evolutionary psychologist relies on little more than her own assessment of plausibility. Nonetheless, I will assume that evolutionary psychology (or something like it) is true—we cognize God with evolution-shaped minds.

So, for the sake of the argument, let us assume that CSR and evolutionary stories of religion are basically on the right path. We think God, as we do all things, with our brain (mind). And we think God, as we do all things, with our very ordinary and natural cognitive faculties (we don't have some special, divinely instilled God-faculty). Those very ordinary cognitive faculties were shaped through mostly unknown evolutionary processes—from the dust, as it were—in response to very specific and species-typical adaptive problems.

Whether or not such dusty religious beliefs are rational is the subject of the remainder of this book.

CHAPTER 3

The Rational Stance

First Pass

We know, or think we do, how beliefs are acquired in the sciences. For example, a scientist accumulates a whole bunch of observation statements, generalizes based on those statements, and infers a natural law. Maybe a scientist watches a lot of falling-down things (apples, say, and raindrops and feathers) and carefully records what she observes; she studies what others who watch a lot of falling-down things have written; and then, with a healthy dose of mathematics, she infers that everything on earth falls down at 9.8 m/sec². Or maybe the scientist develops a theory, the kinetic theory of gases or electron theory, and then devises an experiment or derives the observations that test the theory; the theory is tested to see if those experimental results or observations confirm or disconfirm it. And then it is tested again and again. Any theory that survives one of these very rigorous processes is surely rational.

Must all of our beliefs model those of the scientist in order to be rational? If all of our beliefs are like belief in a scientific theory, then every belief would have to pass science-like tests or experiments.[1]

I don't form beliefs about my son, Evan, because those beliefs (or he) have passed a battery of tests or experiments. I just see him or hear him and find myself believing that he exists and is a person with various thoughts and wishes (and I respond to him as I do to persons and not, say, as I do to porcupines or pine trees). I laugh with him and hope for him and, unlike my relations with porcupines and pine trees, hug him. It seems silly to hold my Evan-beliefs hostage to passing scientific tests.

When I look out the window and see a tree, I don't submit the perceptual belief that I see a tree to a battery of tests; I just see it and believe it. I read in a book that in 1773 some colonists disguised as Mohawk Indians tossed 342 chests of tea into the Boston harbor, and I instantly believe that some colonists disguised as Mohawk Indians tossed 342 chests of tea into the Boston harbor. I don't check the footnotes, seek out eyewitness accounts, or read newspapers from 1773. I just read it in a book and believe it. I watch a documentary on the Holocaust and am repulsed by the vicious and immoral actions of a cruel commandant. I don't pause to reflect on the evidence in support of my moral beliefs that genocide is wrong or cruelty to innocent humans is bad. When I see his actions, moral judgments well up inside me without conscious reflection on reasons. (Now, as I am writing, I stop and reflect, as a philosopher, thinking that we don't have any scientific evidence whatsoever for any of our moral beliefs, including "genocide is wrong" and "cruelty to innocent humans is bad.") I stop someone on the street and ask them for the time; they reply, "12:45," and I thank them, believing that it is now (about) 12:45 p.m. I don't ask anyone else; I don't google "atomic clock" to find a site to confirm the time; I don't check the arc of the sun. A person tells me the time, and I believe them. Perceptual beliefs, beliefs based on authority, moral beliefs, and testimonial beliefs (and a whole bunch more) are rational without having to pass scientific tests.

I will argue in this chapter that sometimes, maybe most of the time, I just find myself believing things (without any reflection whatsoever on evidence or carefully attending to an argument)—and so do you. Rationality, so it seems, cannot be restricted simply to what passes science-like tests.

What then makes our believings acceptable, rational even?

In this chapter, I develop an account of rationality that accords with much of what we have learned about the science of the mind in the past fifty or so years. This view is defended by philosophers (including me).[2] Not all of them, of course; for every philosophical view there is an equal and opposite philosophical view. So I will present, at least for your entertainment, an account of rationality that I think is plausible and defensible, one that appeals to our good sense.

Why Be Rational?

First things first: Why should we even care about being rational? Why not let everyone believe as they please, rationality be damned? I think the reason we should care about being rational is that all or almost all of us are truth-seekers. Most of us want to orient our lives around the truth. From the trivial to the significant, we want to know the truth. We want to know if a famous movie star still loves his wife or is secretly devising a reunion with his former girlfriend; we want the truth about our political candidates and even the right political system; and we wanted to be sure that Iraq had weapons of mass destruction or that Iran intends to build a nuclear weapon (and attack Israel). And if there is a God, let it be known, and let us live our lives accordingly. But if there is no God, then let that be known, and let us get on with our lives in accord with that truth. The goal of rationality or of being rational is to get more in touch with the truth.[3] Seeking to be rational is seeking the truth, which is something we all want. Rationality, then, is highly desirable in our quest to live our lives properly oriented around the truth. And most of us don't want to live a lie.[4]

What If?

What does it mean to be rational? Let's see if we can make some progress on the notion of rationality by way of some stories.

Suppose one day, as you are going through your mail, you find an envelope addressed to you with no return address. Thinking it somewhat peculiar, you open the letter to find a simple message: "Your spouse is cheating on you." No pictures are included, no dates, times, or names—just the assertion of your spouse's unfaithfulness. You have already enjoyed fifteen good and, as far as you know, faithful years with your spouse. His or her behavior has not changed dramatically in the past few years. Except for this allegation, you have no reason to believe that there has been a breach in the relationship. What should you do? Confront him or her with what you now take to be the truth? Hire a detective to follow your spouse for a week and hope against hope that the letter is a hoax? Or simply remain secure in the trust you have been building up all of these years?

Here is another "what-if." Your child, Katie, comes home after taking her first philosophy course in college and completely ignores you, but not in the way she did in high school when you were merely totally annoying; now she ignores you as though you weren't even in the room. When queried, Katie arrogantly explains the so-called problem of other minds, which attempts to answer the question "How do we know that other minds and, by extension, other people exist?" How do we know that other people aren't simply cleverly constructed robots with excellent makeup jobs? How do we know that behind the person facade lies a real person, someone with thoughts, feelings, and desires? Katie takes this problem, the problem of other minds, really seriously. She knows firsthand that *she* is a person because she experiences *her own* thoughts, feelings, and desires.

But she cannot tell just by looking if I am or anyone else is a person. She cannot tell, for example, if *I* am a person because she cannot experience my feelings, think my thoughts, or feel my pain (even noted empath Bill Clinton cannot *really* feel another person's pain). But because these feelings and desires are all essential to being a person, and because she has no access into other people's inner experience, she cannot really know that I or anyone else exists. I protest, but, bored and unsure of my personhood, she sighs and whispers (apparently to herself), "Whatever," and turns back to her video game.

How do you treat someone whose personhood status is pending? Do you hire a philosophical detective to search for a proof that some people-like things really are people? Do you avoid hugging or loving your child in the meantime, given your aversion to cuddling with machines? Or do you simply trust your deep-seated conviction that, in spite of the lack of evidence, your child is a person and deserves to be treated as such?

Suppose you are in a physics class and your hard-nosed, fact-loving, faith-despising professor insists that no one should believe anything without first testing it (and making sure it has passed those tests). "Belief must be based on evidence," he shouts. "Any belief not based on evidence should be rejected." And then he mocks students for their benighted and poorly based (mostly Christian) religious and (mostly Republican) political beliefs. He laughs at people who believe in free will and the objectivity of morality. When a student protests, he shouts her down, demanding of her, "Where is the evidence, where is the evidence?" But being brave (bordering on foolhardy), you ask him why he assigned the class to read a textbook since it requires students to take what the authors say by way of testimony, not careful assessment of evidence. He replies that these people are brilliant, have PhDs in their specialty, and *they* have examined the evidence. "But," you demur, "that means *I* must accept what they say simply on the basis of authority (and not careful assessment of the evidence)." You note that even your esteemed professor cannot be expert in everything—there are too many theories, too many facts, too many fields outside his area of expertise; and even within his area of expertise, he must rely on reports of countless other scientists (he cannot repeat all of their experiments). Even he has to accept what people say on the basis of testimony and authority. You go on to ask about the unproven assumptions of science—the principles of logic and mathematics, the belief that the future will be like the past, the claim that we can infer universal statements from a finite body of data, and the basic reliability of the senses—and you ask, "Don't we all rationally affirm many really important things that aren't or maybe even can't be based on evidence?" Your irritated professor huffs, "We don't have time for all this philosophy crap," derisively snort-laughs, and returns, self-satisfied, to his lecture on entropy.

The Demand for Evidence

If you think it odd to submit your spouse's fidelity or the personhood of your child or the testimony on which scientific teaching or inquiry relies or the assumptions on which science is based to the demand for evidence, you are already sensitive to the excesses of a strict and "scientific" demand for evidence for some of our very significant beliefs. Should or even can we demand adequate evidential support of most of our beliefs?

Before addressing this question, let me speak very briefly about the kind of evidence I have in mind. I restrict evidence, for purposes of our discussion so far, to what philosophers call "propositional evidence." Propositions, also known as statements or declarative sentences, are the primary bearers of truth-value (and so are either true or false). Propositions include, among countless other declarative sentences,

> *Skunks smell stinky.*
> *Beijing is crowded.*
> *2 + 2 = 4.*
> *Rome fell because of lead poisoning.*
> *Salted butter tastes better than unsalted butter.*

The first three propositions are true, the fourth is false, and the last is debatable but, I think, true. The proposition

> *There is an even number of blades of grass in my lawn*

is a proposition (that is, it is either true or false), but none of us is able to determine whether it is true or false. Finally, when we believe something, we affirm (or deny) a proposition. I believe that Michael Jordan is the best basketball player of all time and that raising the minimum wage will not increase unemployment; this means that I affirm the proposition

> *Michael Jordan is the best basketball player of all time*

and I deny the proposition

Raising the minimum wage will increase unemployment.

Of course, as with respect to most beliefs, I could be wrong.

You might have thought, using your ordinary understandings of evidence, that a fingerprint, a fossil, or a warm winter day were evidence of a person's guilt, a new dinosaur, or global warming. Or maybe you thought that seeing faces is evidence of a person's thoughts, feelings, or desires (and therefore that they are persons). Or maybe you thought that logic and math are self-evident (they carry their evidence on their shoulder, as it were; they don't need to be proved for us to see that they are true). I agree that much of this counts or could count as evidence. However, the philosopher's demand for evidence typically restricts evidence to propositions, which, as we have seen, are declarative sentences, which are, in turn, bearers of truth and falsity. I will explain later in the chapter, in the section entitled "The Aladdin Problem," why some philosophers favor propositional evidence.

Those who demand a link between rationality and propositional evidence restrict rationality to beliefs established by argument. Arguments are sets of propositions in which some of them (the premises) are offered in support of another of them (the conclusion).

All people are mortal.
Socrates is a person.
Therefore, Socrates is mortal.

This classic syllogism offers the first two propositions as evidential support for the conclusion. In this argument, the conclusion deductively follows from the premises (note, all of the information in the conclusion is contained already in the premises).

In inductive arguments, the evidence is finite and the conclusion is (in principle) infinite. For example, again a classic:

Observed swan$_1$ is white.
Observed swan$_2$ is white.
Observed swan$_3$ is white. . . .
Observed swan$_n$ is white (where n is a finite number).
Therefore, all swans are white.

In the case of inductive arguments, the inference is essentially ampli-
ative—the (infinite[ish]) conclusion vastly exceeds the (finite) evidence
contained in the premises. I can observe only so many swans. The entire
human race can observe only so many swans. Therefore, our white swan
premises are restricted by human finitude. Yet we made an inference
about all swans.

The existence of other minds, the past, and the belief that the future
will be like the past, on this view, require the support of a good (propositional) argument. One might think we need arguments in such cases
because we cannot experience the past, other people's feelings, or the future. We need something, a good argument, to move beyond what we can
experience to those things we cannot experience (things we nonetheless
believe in).

We might demand evidence for all of our beliefs because we think this is
how science works, and science is the most rational enterprise humans have
embarked on. Science, with its evidential checks and balances, has achieved
rational consensus where philosophy and theology have failed miserably.
If we take science as our model of rationality, so this goes, we will believe
only what is based on evidence. On this view of science, scientific theories
are inferred from observation statements (propositions), not, technically
speaking, observations. From a careful analysis of propositions about, for
example, falling things, Newton induced the law of universal gravitation.
From tables of statements that recorded observations concerning the inter-
relation of heat, gases, and pressure, Boyle affirmed the kinetic theory of
gases. The same holds, or so it is claimed, for natural laws and theories from
Boyle's law to $E = mc^2$. If we want to rationally move from observations
about metal containers filled with gas expanding when heated to belief in
unobservable atoms, we had better have a good argument (based on well-
established propositional evidence). If we want to move beyond our three-
dimensional observed world to rationally affirm Einstein's four-dimensional
curved spacetime, again we had better have a good argument based on well-
established propositional evidence.

Extending this demand for evidence: rational belief in other minds, the
past, and the future requires a good argument to go beyond what we experience to what we cannot experience. Richard Dawkins claims that rational
belief in God's existence must be based on propositional evidence because
God's existence, he claims, is a scientific hypothesis. God's existence is rational only if well supported, like the law of gravity and the kinetic theory of
gases, by good propositional evidence. He writes: "The presence or absence

of a creative super-intelligence is unequivocally a scientific question." If belief in God is a scientific question, it shouldn't be accepted without adequate propositional support.[5]

Are beliefs in other minds, God, and the past really like scientific hypotheses (and so stand in need of the support of well-established evidence for their rationality)? Before we can answer that question, we need a better sense of what a scientific hypothesis is.

A scientific hypothesis is made rational by way of explanation and, in most cases, prediction. Scientific ideas such as $E = mc^2$ and the germ theory of disease are supported by their surprising and illuminating ability to explain and predict (often in mathematically precise ways). The law of universal gravitation both explains the orbits of the planets and the tides and makes precise predictions of the appearances of comets and eclipses.

Most of our beliefs, though, aren't made rational by way of explanation and prediction. I believe that I exist, that I am typing right now, that there is an external world and a past, and that there are other people in the room. I form these beliefs immediately, without reflection or inference, when my cognitive faculties are stimulated in various ways. I don't believe such things because they are the best explanation of this or that sort of experience. If I were to meet you, I would instantly believe that you are a person, not because assuming that you are is the best explanation of my experiential data, but through the functioning of my theory of mind (ToM), which disposes humans to form beliefs about others' mental states.

Such nonscientific beliefs are rational when one's properly functioning cognitive faculties are in the right relationship to the cause of the belief. For example, my perceptual belief that there is a tree in front of me is rational if it is produced when my perceptual faculties are stimulated by said tree. A fond recollection of my mom is rational due to my properly functioning memory (and my theory of mind) and by me being in the right causal relation to the events (events that included my mother) that gave rise to the memory. I don't hold perceptual or memory beliefs because they are the best explanation of my experiences. Most of our beliefs are rational in this way—when one's properly functioning cognitive faculties are in the right relationship to belief.

So perhaps *rational belief does not (always) require the support of evidence or argument.*

Resisting the Demand for Evidence

William K. Clifford, an accomplished nineteenth-century mathematician and physicist, famously claimed: "It is wrong, always and everywhere, for anyone to believe anything on insufficient evidence."[6] In his essay "The Ethics of Belief," Clifford shows the force of his position on religious belief, which he thought did not meet the requisite demand for evidence. Let us examine his claim that everything must be believed only on the basis of sufficient evidence (keeping in mind that, on his account, it is wrong to believe in God in the absence of evidence).

In 1870, on an expedition to observe an eclipse in Italy, Clifford survived a shipwreck off the coast of Sicily. It should come as no surprise, then, that Clifford's critique of belief in God uses a parable of a shipowner who knowingly sends an unseaworthy ship to sea:

> He knew that she was old, and not over-well built at the first; that she had seen many seas and climes, and often needed repairs. Doubts had been suggested to him that possibly she was not seaworthy. These doubts preyed upon his mind and made him unhappy; he thought that perhaps he ought to have her thoroughly overhauled and refitted, even though this should put him to great expense. Before the ship sailed, however, he succeeded in overcoming these melancholy reflections. He said to himself that she had gone safely through so many voyages and weathered so many storms that it was idle to suppose she would not come safely home from this trip also. He would put his trust in Providence, which could hardly fail to protect all these unhappy families that were leaving their fatherland to seek for better times elsewhere. He would dismiss from his mind all ungenerous suspicions about the honesty of builders and contractors. In such ways he acquired a sincere and comfortable conviction that his vessel was thoroughly safe and seaworthy; he watched her departure with a light heart, and benevolent wishes for the success of the exiles in their strange new home that was to be; and he got his insurance-money when she went down in mid-ocean and told no tales.[7]

The crucial point for Clifford does not concern the belief itself, but how the belief was acquired. The shipowner acquired his belief that the ship was seaworthy, not by carefully attending to the evidence (for the evidence was to the contrary), but rather *by suppressing both his doubts and the counterevidence.* The sincerity of his belief was irrelevant to the rightness or wrongness of his believing, because his lack of evidence denied him the right to believe in the ship's seaworthiness. He ought to have acquired the belief by patient inquiry, not by selfishly acceding to his passions. His intentional stifling of both the evidence and his doubts made him fully responsible for the deaths of his passengers. From this parable Clifford draws a general lesson: "*It is wrong, always and everywhere, for anyone to believe anything on insufficient evidence.*"[8]

Clifford's examples powerfully demonstrate that in some cases, like the seaworthiness of ships, beliefs require evidence in order to be rational. And no one would or should disagree that some beliefs require evidence in order to be rational. No one is advocating that no beliefs ever require the support of evidence. But *all* beliefs in *every* circumstance? That is an exceedingly strong claim to make and, it turns out, one not based on evidence.

The first reason to suppose that not all of our beliefs can be based on evidence is the *regress argument*. Consider your belief A (where A is a proposition that you affirm). If A is rational, according to the universal demand for (propositional) evidence, it must be based on some (propositional) evidence (say, B). But if B is rational, it must likewise be based on some evidence (say, C). And if C is rational, it must be based on D, and D on E, and E on F, and so on. If *every* belief must be based on evidence, then having just a single belief would require one to have an infinite regress of beliefs. But none of us, in this busy day and age (or in any age, busy or not), has the mental space or energy or time to hold an infinite number of beliefs. So, if we are capable of rationally believing anything, there must be some beliefs that we can take as evidence but that aren't based on evidence themselves. In other words, we must be able to *start* with some beliefs. There cannot be a *universal* demand for evidence because, to avoid an infinite regress, some beliefs must be accepted and acceptable without the support of evidence.

The beliefs we start with are called, variously, "immediate" (not mediated by other beliefs), "noninferential" (not inferred from other beliefs), or "basic" (not based on other beliefs). Typical basic beliefs include "I see a tree," "I hear the tweet of a bird," "I remember having a good cup of coffee earlier this morning," and "Jon seems angry." When prompted and with the right input, we simply find ourselves with these beliefs (we don't reason to them). Such beliefs constitute the foundation or evidential basis of our knowledge. Our other beliefs are based on (mediated by, inferred from) our foundational beliefs.

The second reason to reject a universal demand for evidence is that *there is no evidence to support the universal demand for evidence*. Consider what Clifford might allow us to take as evidence: beliefs that we acquire through sensory experience—those acquired through seeing, hearing, touching, tasting, or smelling—and beliefs that are self-evident, like logic and mathematics; these two sorts of beliefs are often what some think are the basic beliefs allowable in the sciences.[9] Science, so it is claimed, is grounded in sensory experiences and then is built up through math and logic into natural laws. Can such a Cliffordian conception of evidence support a universal demand for evidence?

Make a list of all of your experiential or sensory beliefs, such as the following:

> *The sky is blue.*
> *Grass is green.*
> *Most trees are taller than grasshoppers.*
> *Honey is sweet.*
> *That smells like a rose.*
> *Slugs leave a slimy trail.*

Now add to this a list of all your logical and mathematical beliefs. Here are some candidates:

> *2 + 2 = 4.*
> *Every proposition is either true or false.*
> *In Euclidean geometry the sum of the interior angles of a triangle equals 180°.*

Take all of these propositions together and try to deduce from them the conclusion that it is wrong, always and everywhere, for anyone to believe anything on insufficient evidence. You'll notice that none of the propositions allowed as evidence have anything at all to do with that conclusion. You can no more deduce the universal demand for evidence from this set of propositions than you can deduce, "Every third step that you take must be accompanied by a loud shriek"; the evidence is irrelevant to both statements. Clifford's universal demand for evidence cannot satisfy its own standard. Therefore, by Clifford's own criterion, it must be irrational.

There are two takeaway lessons from this exercise. First, if there were a universal demand for (propositional) evidence, all of our beliefs would be ruled out as unjustified or irrational (the infinite regress argument). Second, even if we allowed evidence such as logic plus sensory experience, most of our beliefs, those beyond our immediate experience, would be ruled out as irrational. A universal demand for evidence is too demanding.[10]

Belief Begins with Trust

We, finite beings that we are, simply cannot meet a universal demand for evidence. Consider all of the beliefs that you currently hold. How many of those have met this strict demand for evidence? Clifford intends for all of us, like scientists in a laboratory, to test all of our beliefs. Any beliefs that fail the evidential test should be rejected as irrational. Could many of your beliefs survive such tests? Consider how few of your beliefs actually meet Clifford's evidential standard.

Think of how many of your beliefs, even scientific ones, you have acquired *just because someone told you*. Not having been to Paraguay, I have only testimonial evidence that Paraguay is a country in South America. Moreover, perhaps mapmakers have conspired to delude us about the existence of Paraguay. Since I have been to relatively few countries around the world, I must believe in the existence of most countries (and that other people inhabit them and speak in the language of that country) without the support of adequate evidence. Even if, while in a foreign country with lots of people speaking Spanish, I were to look around and see Paraguayan signposts ("Bienvenidos a Paraguay"), I would still have to trust that everyone around was not deceiving me.

I believe that $E = mc^2$ and that matter is made up of tiny little particles (or waves or wave-particles), not because of experiments in a chemistry or physics lab (all of my experiments failed anyway), but because my science teachers and scientific textbooks told me so. But surely it is rational, in many cases, to believe what others tell us.

Many of the beliefs that I have acquired—including most of my scientific ones—are based on my trust in my teachers and textbooks and on other relevant authorities, not on careful consideration of what Clifford would consider adequate evidence.

In a large number of cases, this demand for evidence simply cannot be met with the cognitive equipment that we have.

Take the *problem of other minds* mentioned in the what-if story at the beginning of this chapter. I don't have experiential access to the inner life of another person—her thoughts, feelings, or desires. I have direct access only to faces and bodies, not to what goes on inside a person's "mind." So I couldn't base my belief in other persons on the only evidence that is relevant to establishing personhood. Since thoughts, feelings, and desires are essential to being a person, lacking access to the thoughts and feelings of another person means that I lack access to the only evidence that could establish personhood. I know that *I* am a person because I do have access to my thoughts, feelings, and desires. But that cannot be how I know that anyone else is a person (since I don't have access to their thoughts, feelings, and desires). And I cannot infer from my own experiences, on any scientific grounds or logical grounds, anyone else's experiences. There is simply no good argument for other persons. And yet, believing in other persons is perfectly rational (and denying the existence of other persons—or thinking them merely robots—seems crazy).

In addition to other persons, no one has been able to prove that there is an *external world*—a world outside one's own mind or independent of one's own experiences. Most of us firmly believe that there is a world independent of our minds, one that causes our sensations of trees, say, and mountains and stars. We aren't, we believe, trapped in a world of our own imaginings. But if my only evidence is my own experience, how can I prove that there is something outside of my own experience?

Moreover, no one has ever been able to prove that we were not created five minutes ago with a full set of memories or that *the past* actually happened or that *the future will resemble the past* (in the scientific sense). I don't have any experiences of the past or future—my experiences are all now (in the present). I see something now, and as soon as I close my eyes I stop having that visual experience, turning it instantly into a present memory (of the past). Given the presentness of all my experiences, experience is poor evidence for the decidedly non-present past and future.

If I cannot have rational beliefs about the past and future, I couldn't possibly acquire any *scientific knowledge* because scientific theories or natural laws are typically about how things behave in all places at all times—past, present, and future. Moreover, modern science assumes such unprovable things as the existence of the external world, other persons, the past, and that the future will be like the past. Without an external world and the collective wisdom of other inquirers, and without the principle of induction (that the future will be like the past), science would be impossible.

The list could go on and on. There is a limit to the things that human beings can base on evidence. A great deal of what we believe is based on trust, even in the sciences, and not on evidence or arguments.

While I use the term "trust" here, I don't oppose trust to knowledge. For surely we know (though on trust because we cannot prove) that other persons exist, the earth is more than five minutes old, the sun will rise tomorrow, Paul converted to Christianity, and dinosaurs once roamed the earth. We know lots of things that we cannot prove; we know them, without proof. And we know them *through our cognitive faculties that (noninferentially) produce such beliefs.* We rely on our memory to produce memory beliefs (I remember having coffee this morning). We rely on an inductive faculty to produce beliefs about the veracity of natural laws (if you were to let go of this book, gravity would cause it to fall to the ground). We rely on our cognitive faculties when we believe that there are other persons, a past, and a world independent of our mind and when believing what other people tell us. We cannot help but trust our cognitive dispositions. And beliefs thusly produced are rational.

Descartes's Failure

In the eighteenth century, Thomas Reid, a Scottish philosopher whose ideas have recently enjoyed increased attention, defended the wide variety of beliefs that our minds naturally produce. He developed a very powerful conception of human knowledge that affirms our reliance upon cognitive faculties and rejects the Enlightenment's universal demand for evidence. Reid had the distinction of succeeding Adam Smith, the father of free-market capitalism, as professor of moral philosophy at the University of Glasgow; Reid's success was both temporal and, ironically, fiscal (the more popular Reid taught many more students than Smith under a system in which professors were paid per pupil!). The following defense of reason is not precisely Reid's defense; rather, it is in the spirit of Reid. This Reidian defense of rationality endorses most beliefs of most ordinary believers and, unlike much of philosophy, has a great deal of plausibility.

Reid blames the universal demand for evidence, which he thinks leads to skepticism, on René Descartes (1596–1650). Descartes, as every "Introduction to Philosophy" student learns, began by doubting everything that could be doubted. Descartes keenly desired certainty and sought it by eliminating the uncertain (whatever can be doubted). Descartes's so-called method of doubt involved rejecting as false anything that could possibly be doubted until he could discover something for which there were no possible doubts—the sure, certain foundation on which to build knowledge. For instance, when Descartes considered the beliefs he had acquired on the basis of his senses (seeing, hearing, and so on), he realized that sometimes his senses deceived him. Therefore, there was a reason to doubt his senses; as such, his senses could not provide him with the source of certainty that he desired.

Descartes proceeded from doubting his senses to doubting the external world, and from there to doubting even mathematical principles (like 1 + 1 = 2). His argument involved a merely possible evil demon who could be deceiving us even about arithmetic (he could make such statements seem certain all the while they were false). The only thing Descartes could be absolutely certain of was his own existence (not even that he had a body—because knowing that he had a body would require the use of his [deceitful] senses). From his own existence, Descartes attempted to build up knowledge of himself, the world, the past, the present, and the future. His only tool was logic, which I will call *reasoning to*. We reason, using logic, to a belief from other beliefs that we already hold; but, Reid argues, Descartes left himself with precious little to reason *from*.

If Descartes were to restrict himself to what strictly and logically follows from his own existence, all he could reason to was his own existence! So he failed in his quest to reason to the external world from himself and his own internal experiences.

On its own, *reasoning to* is impotent in the production of beliefs; one must have some beliefs to reason *from* to reason to something else. And the things Descartes permitted himself to reason from don't provide adequate data for reasoning to the external world. Descartes's failure was repeated over and over by subsequent post-Enlightenment philosophers.

Evidential conceptions of rational belief are defective because our experiential input (which is present moment, finite, fleeting experience) is insufficient to support our belief/knowledge output: the world (which is past, present, future, enduring, and so on). In other words, we have minimal experiential input and massive informational output. Our own experience provides such limited information that it is incapable of supporting our knowledge of the *world*. Even if we were to use logic and mathematics to order our experience and make deductions from it, the world presented to us in our finite experience thus ordered would pale in comparison to the deep and vast world in which we believe.

Think of the world: it extends into the distant past and will proceed into the unforeseen future; its physical dimensions are both inconceivably vast and tiny; it includes people, some of whom lived long ago, some who live in the present, and some who will live in the future; it includes me, a conscious and self-conscious person, who persists through time and recalls finishing first in a high school relay (thanks to three much-faster teammates). Now think of your own sensory experiences. Could they, when supplemented with the rules of logic and mathematics, produce the world (or, more precisely, produce justified beliefs about the world)? Even if we were to add the experiences of others to our repository of information, we would still be incapable of deriving the world with our joint experience and logic.

If the universal demand for reason (basing everything on sufficient evidence) leads to an abstract, arid, and tiny world—a "world" that includes only me and my experiences—we should abandon the Enlightenment understanding of reason. The reliance upon sufficient proof from very finite experience has led us astray. It is time to find a new path. Precious little could be proved by Enlightenment philosophers, and what ought to be rejected, says Reid, aren't our ordinary beliefs but the universal demand for proof. You cannot prove everything. Moreover, you don't need to prove everything.

Cognitive Dispositions

Fortunately, as noted in the previous chapter, we come equipped with further inbuilt cognitive faculties or dispositions or inclinations that produce, where experience and logic fail, substantial beliefs about the world. We have many cognitive dispositions, not just reasoning, that produce beliefs about the world. They fill up our repository of belief where logic and experience could not. Without this further inbuilt cognitive equipment, we would have precious few significant, rational beliefs—no external world, no past, no science, etc. In this section, I will draw upon the insights of Reid, extending his views through recent work in cognitive science and illustrating the relevant cognitive faculties with contemporary illustrations.

We have cognitive faculties of *sense* and *memory*. Unlike Descartes, who distrusted his five senses, Reid contends that our sensory beliefs are rational even if they occasionally produce mistaken beliefs. And just as it is rational for me now to believe that Rebecca is wearing a black sweater (while I am looking at her and while she is wearing a black sweater), so, too, it will be rational for me to believe later in the day that she was wearing a black sweater based on my memory. Since the belief-producing faculties of perception and memory are as much a part of the human constitution as is reasoning, there is no reason to exalt reasoning over sense and memory.

Another cognitive faculty that we enjoy produces belief in an enduring self. The philosophical concept of a self is complex and tricky. While you might think that you are the same person you were in Miss Schwartz's third-grade class, proving that you are the same person from then until now has proven impossible. Lucky for us, we have a cognitive faculty that produces *belief in a self*, that persists through time, and that unifies us and our experiences. Although we have a strong tendency to believe that we have an enduring self, no one has offered such a proof. My sensations, which are discrete, are in need of an enduring self to unify them into, well, me (myself). But we cannot infer an enduring self from our discrete sense experiences. So, if the belief in a self must be based on an argument using only one's experience as evidence (remember, memories don't count), then belief in a self is folly.

Imagine a murder trial in which the defendant, let us call him "OJ," claims that he didn't commit the murder because he had no self that endured through time. There were, a long time ago, anger sensations and stabbing sensations (and being stabbed sensations); and there are currently being-in-trial sensations (the courtroom feels hot, the desk feels smooth, etc.). But OJ alleges in his defense that there is no philosophical theory that proves that the murder sensations are united with the trial sensations by an OJ self that has persisted through time (from the murder to the trial); he, OJ concludes, did not commit the crime (because there is no "he"). OJ might get off by reason of insanity but not by virtue of good reasoning. Even if we cannot prove the existence of an enduring self, it does not follow that we don't have a self or that belief in a self that persists through time is irrational.

Our cognitive faculties also produce *belief in the past*, which is assumed in every historical belief. For example, consider the beliefs that Caesar crossed the Rubicon and that the Chinese invented gunpowder. Since I cannot experience the past directly, my beliefs concerning Caesar and the inventor of gunpowder aren't based on any sensations of Caesar or of an ancient Chinese inventor. Imagine a history student who, upon taking her final examination, turns in a single statement that says, "Because no one could have any experiential contact with the past, it is not rational to believe in any past events or even the past at all." While offering a creative excuse, her failure to use her nonsensory cognitive faculties would earn her an "F" in her class.

The general belief that the past exists is related to our memory, which produces beliefs directly and immediately by our cognitive faculties (given the right sorts of prompting or circumstances). For example, if I were asked, "What did you have for breakfast this morning?" this question would trigger a memory, which would automatically produce a memory belief—"I had oatmeal." I have no current sensations of oatmeal—I don't see, hear, taste, or smell it. I don't infer the belief that I had oatmeal for breakfast from any of my current sensations; rather, I just immediately form the belief upon being prompted by the question. We trust our memory as (more or less) reliable even though we know that it sometimes fails us.

Even within the domain of science, the redoubtable sphere of experiential and experimental rigor, *the uniformity of nature* (that the future will be like the past) is simply assumed. Science makes universal generalizations based on a finite set of extremely limited experiences. We cannot experience most of the universe, and, moreover, even the future exceeds our puny experiential grasp. How can we make rational judgments about things we cannot see or fully understand? The law of universal gravitation, for example, states that every object in the universe is attracted to every other object in the universe in direct proportion to their masses but in indirect proportion to their distance. This is supposed to be true of every two objects, everywhere in the universe, past, present, and future. But we can see the behavior of only the tiniest fraction of objects in the universe. We can pile finite experiences on top of finite experiences day and night, but we will never be able to generalize to every object everywhere without assuming the uniformity of nature. The practice of science would be impossible without our natural cognitive ability to generalize from a finite set of data to everything, past, present, and future.

So we have a tendency or faculty to believe, in the appropriate circumstances, that there is an external world, that we have an enduring self, that there are other persons, that our memories are reliable, that the future will be like the past, and that nature is uniform. The significance of the cognitive faculties that produce these beliefs is that, with the exception of the reasoning disposition, they produce their effects immediately, without the evidential support of other beliefs. Most of our cognitive faculties produce beliefs in us in an immediate, noninferential manner. And we don't typically seek to justify these sorts of beliefs by reasoning to them. We simply believe them. And, I'll argue, we are perfectly rational in accepting them.

We have no other option: our cognitive faculties are all we have to work with in belief acquisition and understanding the world. These are all the tools we got.

Not all of our beliefs are immediate. Some beliefs are acquired and maintained because of other beliefs we hold, by *reasoning to* them. Sometimes scientific theories—for instance, the belief that there are electrons or that $E = mc^2$—are acquired by performing certain experiments in a laboratory or examining the observational evidence. A scientist, upon considering the mathematical formulas relating the orbits of the planets, may acquire the belief in the universal law of gravitation. Also we can imagine many everyday beliefs acquired through reasoning. After hearing testimony at a trial, one might infer that the defendant is guilty. Likewise, after weighing the nutritional evidence, one may believe that giving up bacon will reduce one's cholesterol count. These are examples of beliefs generated from the reasoning disposition.

Yet even the honest scientist must confess a dependence on inferential principles, principles that scientists simply trust, such as induction. Given our essentially finite number of human experiences (put into propositions), the probability of any universal statement (including every natural law or theory) would be zero. Without trusting in the inductive principle, science would be impossible. With it, stand back!

Still, the vast majority of beliefs that we hold aren't acquired by reasoning to them but are produced immediately and noninferentially by our various cognitive dispositions.

If we see or hear something, or if our attention is called to something, we immediately form a belief. Someone speaks to us, and we respond to her as a person (without inferring that she is a person). Our very reasoning assumes the unproven validity of logic, and our scientific reasoning assumes the unproven uniformity of nature. And, if we are honest, a huge proportion of what we believe is acquired simply because someone told us, whether another person, teacher, newspaper, or magazine.

We should marvel, I think, at our astounding inbuilt cognitive mechanisms. Without them, we would be lonely, solitary selves, existing only now. With them, we get other persons, the past, the future, a self, the external world, and the principles that undergird science. Add to them our astonishing ability to reason, to think counterfactually, and to think mathematically and creatively and you get the whole of modern science. We are deeply finite—that is, deeply dependent on our inbuilt cognitive equipment to know much of anything at all. But from it, we know about atoms, distant galaxies, dinosaurs, and Neanderthals. Awe seems the right response to our abilities to grasp the world.

Believing What Others Tell Us

The cognitive faculty that governs our acceptance of what others tell us, sometimes called the *principle of testimony*, is a profound and extensive source of human beliefs. From believing the stranger's response to an inquiry about the time of day, to accepting what your teachers told you in class or what you read in a book, the reliance upon *others* for our beliefs is deep and pervasive. Even in science—where evidence and proofs are sought—most scientific beliefs are based on what others have said. I said previously that one might affirm the universal law of gravitation only after carefully weighing the evidence. But most of us nonphysicists can scarcely understand the universal law of gravitation, much less weigh and measure the evidence in its favor. And even if I had tried the experiment, my data would probably have given me evidence that $E = mc^3$ or $E = mc^{2.24876629}$ (surely not $E = mc^2$). I believe that $E = mc^2$ because brilliant people have amassed the evidence and I have simply accepted what they told me.

It might be unfitting for a physicist to accept some scientific hypotheses—say, some of them in their area of speciality—without carefully considering the evidence. The physicist must carefully attend to the evidence and to the counterevidence and not assent to some (but few) scientific hypotheses without the support of the evidence. We would not take seriously a scientist who committed herself to a scientific hypothesis because it just occurred to her on a whim or because it "just seemed to be true." Yet even the best and most fastidious scientist must accept a great deal of what other scientists tell her. The physicist, incapable of proving every theory, must accept some theories' experimental results and observations on the authority of her peers. As a nonphysicist, I believe that there are electrons, that the sun is made mostly of hydrogen, and that black holes exist only because scientists have told me.

I also believe there is a country called Uzbekistan and that it has a population of about 28,000,000, that the best coffee is Jamaica Blue Mountain (and from Jamaica), and that the largest ocean is the Pacific because others have told me. Even if I were to try to confirm these matters, I would have to rely on what others told me (mapmakers, census counters, historians, importers, salespeople, museum guides, and so on).

Reid claims (incorrectly) that the willingness to believe what others tell us is unlimited in children, because children accept without question whatever anyone tells them. But as they grow and mature, children begin to question the testimony of others. They begin to ask questions about what others tell them, in part because what they have been told sometimes conflicts with other things they have been told. They begin to realize the truth that beliefs produced by testimony aren't infallible. When such beliefs come into conflict, one must call upon one or another of one's cognitive faculties to resolve the conflict.

Consider an example. Suppose Will tells you that he saw Gabe walking along the river arm in arm with Alicia last Tuesday night. But when you see Gabe and ask him about Alicia, he denies that he was with Alicia on Tuesday. Now you have two beliefs vying for your acceptance: "Gabe was with Alicia" and "Gabe was not with Alicia." Which should you believe? You might just suspend belief, or you might ask someone else, or you might check out the security videos posted along the river. Suppose you check those cameras and see that Gabe was not with Alicia. While you were initially rational in believing what Will told you, because it is generally fine to believe what others tell us, Gabe's denial causes a conflict of testimony. You set about to resolve the conflict by further engaging your cognitive faculties (testimony or visual). Suppose that the video cameras undermine or defeat Will's testimony; if so, they undermine or defeat the rationality of your originally rational belief that Gabe was walking with Alicia. Your original belief was initially rational, but its rationality was initially called into question by Gabe's testimony and subsequently defeated by your checking the videos.

Although people sometimes lie to us and sometimes people give us quite different reports of what they saw, most people most of the time tell the truth. And we simply cannot live without relying on the testimony of others. Since there is too much to know, we are right to believe what others tell us. For example, while I know a few areas of philosophy, I rely on others to tell me about Bergson, Han Feizi, and Vasistha. And as a philosopher I must rely on my biologist friends and texts to explain alleles, epigenetics, and genetic drift to me (and, similarly, on anthropologists, psychologists, geologists, and economists to supply other information). I read newspapers in which reporters tell me what's happened in Jakarta, Oklahoma, and Osaka (and I believe what they tell me). I call my carpenter brother, Jon, to ask advice on carpentry, electronics, and cars (and I take his advice without confirming it). In short, testimony is essential and, for the most part, trustworthy.

In summary, most of our cognitive faculties produce beliefs immediately, without recourse to argument. Sometimes our beliefs thus produced are in conflict, and we need to consider which of the conflicting beliefs should be accepted and which rejected. Can we develop these insights into a theory of rationality?

Innocent Until Proven Guilty

What does it mean, then, for creatures like us, finite creatures dependent on our cognitive dispositions, to be rational? The recognition of our many cognitive faculties that produce beliefs immediately, without the support of evidence or argument, marks a radical point of departure from the Descartes-influenced Enlightenment conception of rationality, with its universal demand for evidence or argument. Descartes's strategy, "Doubt first, believe second," is simply not tenable for beings like us. Given our cognitive dispositions, we cannot treat all or even most of our beliefs as suspicious or, to borrow and reverse a legal phrase, *guilty until proven innocent*. If we were suspicious of or distrusted either our beliefs or our cognitive dispositions, we would end up believing very little. Following Descartes and Enlightenment principles, we should all be skeptics—disbelievers in the existence of other people, the external world, the past, and the future. For all those who wish to avoid skepticism, read on.

The Reidian understanding of rationality, which I will develop, is more natural for creatures that are finite, limited, dependent, and fallible. We aren't gods—we don't have infallible and indubitable access to the world, and we lack infallible reasoning abilities and principles. Yet our cognitive equipment works fairly well in helping us grasp reality. We have to rely on what we have been equipped with. We can do no other.

Instead of assuming our beliefs are guilty until proven innocent, perhaps, as Reid argues, belief begins with trust. Generally, we can trust our cognitive dispositions, and we ought to trust a belief given to us by our cognitive faculties unless we have substantial grounds for questioning that belief. William Lycan's principle of credulity captures Reid's thought here: "Accept at the outset each of those things that seem to be true."[11]

According to Reid, our initial and rational approach to belief is one of trust, not one of doubt or suspicion. Therefore,

Beliefs are innocent until proven guilty

rather than vice versa. Under this presumption of innocence, a belief ought to be accepted as rational unless or until it is shown to be specious.

Contemporary philosopher Nicholas Wolterstorff affirms Reid's intuitions and develops them into a criterion of rationality:

Reidian Rationality
A person is rationally justified in believing a proposition that is produced
by her cognitive faculties in the appropriate circumstances unless or until
she has adequate reason to cease from believing it.

On this conception of rationality, beliefs produced by our cognitive faculties are rational unless or until one has good reason to cease believing them. That is, we can trust beliefs produced by our cognitive faculties until that belief is undermined or defeated by stronger or better-corroborated beliefs. Our beliefs are innocent until proven guilty, not guilty until proven innocent.[12]

The Aladdin Problem

Reidian rationality is, I think, as good as it gets for finite human beings. As I wrote earlier, we have no other option: our cognitive faculties are all we have to work with in understanding the world. These are the only tools we get.

The human believing condition generates what I'll call "the Aladdin problem." In the 1992 Disney film *Aladdin*, Aladdin rubs a small lamp and, in a puff of smoke and fire, the genie is released. After the verbose genie insults Aladdin a few times, Aladdin attributes his experiences to a trick of the brain: "I must have hit my head harder than I thought." From such modest lamps we might expect oil or water or perhaps, in the Agraba desert, sand. Not genies. Rub a small brass lamp and maybe oil will spill out but never a person (let alone one who can grant your every wish; OK, just three wishes). When Aladdin expresses his incredulity, the genie notes the incongruity: "Phenomenal cosmic powers . . . itty-bitty living space." Massive output, modest input.

The cognitive Aladdin problem is my name for the so-called "poverty of stimulus": itty-bitty experiential input, massive belief output. In a nutshell, the Aladdin problem is that our experiential input (which is present moment, finite, fleeting) is woefully inadequate to generate on its own our belief/knowledge output: the world (past, present, future, enduring, other persons, etc.). I have minimal experiential input—the taste of this and that, these smells and not countless others (some good, some bad), only so many feels (of rough, say, and cold), and I can see only so many swans and falling apples. From my very finite experiential input, I have a massive informational output: I believe in a world that extends vastly beyond my own experience, one that extends into the distant past and will extend according to natural law into the indefinite future. I believe, without possibly being able to experience it, that Emperor Qin killed a lot of innocent people in 225 BCE and that Halley's comet will come within eyeshot of the earth in 2061 CE.

Even if I were to add your and every other person's experiences to my pile of evidence, and then use logic and mathematics to order this pile of experiences, this very finite pile of propositional evidence pales in comparison to the infinitely rich and vast world I believe in. Even putting all human experiences into a set of propositional evidence, this paucity of information is logically incapable of generating our knowledge of the world. The world extends into the distant past and proceeds into the unseen future; its physical dimensions are both inconceivably vast and microscopically tiny; and it includes people, some of whom lived long ago and far away. Our puny set of experiential data, even when supplemented with the rules of logic and mathematics, is incapable of generating beliefs about the world that science and common experience reveal. Yet we rationally believe, where experience and logic alone must fail, in the existence of a world independent of our own finite experiences, a world billions and billions of years old, a world containing atoms and governed by gravity and entropy, a world that will go on as it does long after I die, well into the heat death of the cosmos.

Itty-bitty cognitive input, massive belief output.

The Reidian solution to the cognitive Aladdin problem permits us to trust that our cognitive equipment is up to the task of thinking truly of the world that lies beyond my finite human experience. We simply must concede and accept that we are equipped with cognitive faculties that contribute substantially to our beliefs about the world.[13] And then we must use those dispositions, as best we can, to understand the world.

The Rational Stance

We opened this chapter noting that rationality aims at the truth. While we may aim at the truth, reason does not always hit that target. Surely it was rational for most people in 200 BCE to believe both that the earth was flat and that the earth was at the center of the universe (both are reasonable for some people today in some undeveloped parts of the world). After all, we don't see the earth's curvature, and we don't feel the earth rotating at roughly a thousand miles per hour, nor do we feel as if we are hurtling through space in orbit around the sun at 67,000 miles per hour. Our cognitive faculties can produce and have produced beliefs that one could rationally hold but that are false—beliefs about the place of the earth in the cosmos, the shape of the earth, and the immobility of our planet. Our perceptions have misled and continue to mislead us about the nature of physical reality. It was only as the evidence mounted against geocentrism that such a long and deeply held commonsensical view was abandoned in favor of heliocentrism. If we want to get the truth, then, we must be open to evidence that opposes our otherwise rationally held beliefs.

Evidence is, after all, truth-conducive or truth-indicative (to which modern science is a testament). Evidence is helpful in establishing the truth in a host of areas of inquiry, such as history, court cases, and cooking. Most Chinese (and many of us) believe, based on testimony, that Confucius wrote the *Analects* in roughly 400 BCE. The historical evidence, however, shows that Confucius did not write the *Analects* (though the work alleges to report sayings of the Master), that the received text of the *Analects* is a compilation dated to roughly 150 BCE, and that many of its passages are much later accretions (much later than the time of Confucius). A jurist's initial conviction about the guilt of the accused ("Why would he have been arrested," the jurist thinks to himself at the outset, "if he's not guilty? Sheesh, he sure looks like a criminal") gives way to the truth about his innocence with a careful presentation of the evidence by his attorney. A good chef may learn how to cook a tastier cheese soufflé by trying out lots of new recipes or learning the chemistry of cheese. If we are committed to the truth, we should, at least in some cases, seek supporting evidence.

So we need to amend our understanding of rationality. If we want to discover the truth, we need to be sensitive to the accumulation of evidence and counterevidence. Indeed, acquiring evidence is one of the best guides to truth. However, for all the reasons stated above, acquiring sufficient evidence may not, by itself, help us in the discovery of truth.

We have a lamentable tendency, called confirmation bias, to favor evidence that supports and to ignore evidence that opposes our cherished beliefs. We aren't dispassionate assessors of evidence and counterevidence, acquiring beliefs only after careful objective analysis. We are highly selective attenders to evidence and ignorers of counterevidence, again in ways that preserve our preconceived notions. Confirmation bias, for truth-seekers, is as deeply troubling as it is pervasive. As Raymond Nickerson writes:

> If one were to attempt to identify a single problematic aspect of human reasoning that deserves attention above all others, the confirmation bias would have to be among the candidates for consideration. Many have written about this bias, and it appears to be sufficiently strong and pervasive that one is led to wonder whether the bias, by itself, might account for a significant fraction of the disputes, altercations, and misunderstandings that occur among individuals, groups, and nations.[14]

Confirmation bias is well supported experimentally and is easily recognized in, for example, differing assessments of a presidential candidates' debate. While a Republican and a Democrat may watch the same debate on the same channel and in the same room, they often reach completely different assessments of the "winner," assessments that align perfectly with their previously held political commitments. The Republican highlights this good point and ignores that bad one, whereas the Democrat takes an entirely different set of points to be decisive (failing to even mention, let alone concede, any of the good points favored by her Republican friend). Valdis Krebs found that during the 2008 presidential campaign, those who disliked Barack Obama purchased books that made him look bad and those who liked Obama purchased books in which he looked good.

Confirmation bias suggests that we actively seek out evidence in support of our beliefs and passively avoid evidence that contradicts them. But while we might feel good—after all, we have taken the time to seek out evidence and we have found evidence that bolsters our original convictions—a one-sided body of evidence, one that does not include any counterevidence, is hardly the stuff of truth.

Moreover, evidence against a belief may be more truth-conducive, in some cases, than evidence in favor of a belief. Suppose you are a juror who hears from the prosecutor that the defendant had, as is prized in a criminal case, motive, means, and opportunity. Over the course of several days, the prosecutor assembles evidence in favor of each of these elements. The urge to convict is nearly overwhelming. Then suppose the defense attorney presents irrefutable evidence that the defendant was in Remus, Michigan, at the time of the crime (which was committed in Romulus, Michigan, 170 miles away). In this case, a single case of contradictory evidence swamps an otherwise overwhelming body of supporting evidence. If we are truth-seekers, then, we need to be open to new evidence and willing to revise our beliefs as a result of that new evidence.

We cannot help but rely on or trust our cognitive faculties and the beliefs they (mostly immediately) produce. This is where we start in all human inquiry, and so it is where we must start in our understanding of rationality. Belief begins with trust, pure and simple. But, insofar as we are truth-seekers, we should, if and when we can, both seek supporting evidence and carefully assess evidence contrary to our initially held beliefs. Because following this course of action will assist our quest for the truth, let us propose the rational stance:

The Rational Stance
Trust the beliefs produced immediately by your cognitive faculties,
seek supporting evidence, and open yourself to contrary evidence.

The rational stance captures the insight that belief begins with trust, and it also captures the truth-aimed goal of rationality by seeking supporting evidence and being open to counterevidence.[15]

Before you think the rational stance would or should turn each person into a philosopher (or a scientist), constantly assessing evidence and adjusting one's convictions accordingly, let me offer some cautionary notes. First, in this busy day and age (OK, in any age), most of us simply don't have the time to assess the evidence, pro and con, for most of our beliefs. Most of us, most of the time, in most circumstances, will be rational simply by virtue of accepting beliefs directly delivered by our cognitive dispositions. The scientist in her laboratory has special, role-specific obligations to carefully attend to and assess, in very precise ways, evidence in support of theories. Many others have role-specific obligations to carefully assess evidence as well—for example, historians when writing books or jurists during a trial. But most of us, most of the time, in most of our roles, don't.

Even the scientist in her laboratory judges against a background of un-questioned assumptions about the reliability of her senses, the existence of an external world, the uniformity of nature, and the nature of testimony (since she cannot repeat every experiment for every theory, she must simply accept on the basis of testimony nearly all of her background scientific beliefs). When the scientist leaves her laboratory she will rationally believe, simply trusting her cognitive faculties (and without considering any evidence whatsoever), that her partner is a person, that there is a tree in her backyard, and that honey tastes sweet. She may likewise rationally believe, without evidence or argument, that murder is bad, that she freely chose vanilla instead of chocolate ice cream, and that humans have a natural right to happiness.

Conclusion

Lest we forget, we entered this long journey because we were concerned with the rationality of religious belief. More precisely, we began our journey because many have claimed that belief in God is irrational because it lacks, or is claimed to lack, sufficient evidence. We have given reason to reject the allegedly scientific assumption that all or even most of our beliefs require the support of evidence in order to be rational. And we have offered an alternate conception of rationality, one more suited to the actual cognitive faculties with which we have been equipped. Belief, for us, begins with trust. Is there any reason to think, given this Reidian conception of rationality, that a person could rationally believe in the existence of God?

CHAPTER 4

Reason and Belief in God

The Sensus Divinitatis *and the God-Faculty*

In the previous chapter we spent a great deal of time motivating and defending a theory of rationality within which, as seems right, most of our rational beliefs are produced immediately by various cognitive faculties. We rejected the claim that all beliefs must be based on explicitly formalized arguments in order to be rational. While a universal demand for arguments is seductive, it must finally be resisted because it is inapplicable to our believing condition: we just aren't built (cognitively) to base all of our beliefs on evidence. In fact, we are built (cognitively) to base very few of our beliefs on evidence. Most of our inbuilt information processors produce beliefs without the aid of inference. We see a tree in someone's yard, and our perceptual faculties immediately produce the belief "There is a tree in that yard." You ask what I had for breakfast, and my memory faculties immediately produce the belief "I had toast with jam for breakfast this morning." I hear a bird chirping outside (but don't see it), and my hearing faculties immediately produce the belief that there is a bird just outside my window.

Moreover, just as I don't decide to breathe, I don't decide (on the basis of argument), in the vast majority of cases, what to believe. My cognitive faculties, with the right sort of promptings and experiences, produce most beliefs immediately, without the aid of an argument. With respect to breathing and believing, my faculties do most of the work for me. In most cases, there is no need for me to decide to breathe or believe.

While some beliefs—scientific, for example, or judicial—should be based on evidence (in principle, of course; most of the time most of us accept what scientists tell us on the basis of testimony, not good evidence), most need not be based on evidence. The sorts of beliefs that we do and must *reason to* are a considerably smaller set of beliefs than the set of beliefs we do and must accept *without* the support of evidence. That is the long and short of the human believing condition. In most cases, we must simply rely on or trust our intellectual equipment to produce beliefs, without propositional evidence or argument, in the appropriate circumstances. Even if we were to believe on the basis of argument, we would still have to trust a host of cognitive faculties, including our reasoning faculty. Belief, we said, begins with trust, trust both in our cognitive faculties and in the beliefs produced by them.

Putting this all together, we argued that beliefs are innocent (rational) until proven guilty. It is perfectly rational, we argued, to hold a belief produced by our cognitive faculties unless or until it (or the cognitive faculty that produced it) is shown to be specious.

We elaborated this claim into

The Rational Stance
Trust the beliefs produced immediately by your cognitive faculties,
seek supporting evidence, and open yourself to contrary evidence.

What would happen if we took these insights and applied the rational stance to belief in God?

The rational stance is rooted in what I call the Reid-Wolterstorff-Plantinga view. As we shall see, the application of the Reid-Wolterstorff-Plantinga view to religious belief has been called Reformed epistemology (from the Reformation theologian John Calvin). Since this view also finds roots in Anselm, Augustine, Aquinas, Bonaventure, and a host of other theologians and philosophers, calling it Reformed epistemology mislocates the project. That said, calling it the Reid-Wolterstorff-Plantinga-Calvin-Anselm-Augustine-Aquinas-Bonaventure view might be more accurate but less punchy.

Along with our memory and perceptual faculties, the cognitive science of religion suggests that we have a cognitive faculty that produces God-beliefs without recourse to argument. Is belief in God somehow different from our other immediately produced beliefs or do we have a cognitive faculty that immediately produces belief in God?

As we have in previous chapters, let us begin with a couple of what-ifs.

Suppose that you believe that there is a God because your parents taught you from the cradle up that God exists. As far as you can remember, you have always believed in God. You cannot recall a time in your life when God's existence didn't seem as obvious as the existence of other persons. God's nonexistence has never occurred to you.

Or suppose that during a mountain-top retreat you are suddenly over-whelmed by the feeling that God created the universe and that God created you and loves you. You begin to believe in God, but not because you are persuaded by an argument; you just start believing in God, finding your-self believing what you had previously denied. You have not ignored any arguments; you just have not attended to them. Arguments were not a factor in your newfound faith. Theistic arguments simply never occurred to you.[1]

These fairly typical human experiences,[2] the ubiquity of religious be-lief, and studies in the cognitive science of religion suggest that we are dis-posed to belief in God rather easily, naturally, and noninferentially.

The theological motivation to claim that we have been endowed with a God-faculty seems clear: if a loving God created human beings and wanted them to be in relation to him, then God would have outfitted us with the cognitive equipment we needed to love and know him; and if God is a person, as is widely believed, then the ways of knowing God will paral-lel the ways we know other (human) persons. The Christian tradition, rooted as it is in the Bible, takes its inspiration on this matter from Paul, who writes: "Since the creation of the world God's invisible qualities—his eternal power and divine nature—have been clearly seen, being understood from what has been made" (Rom. 1:20). This passage has inspired some Christian theologians to defend a kind of natural, basic, almost instinctual knowledge of God.

Reformed epistemology holds that one may properly and rationally be-lieve in God's existence without basing one's belief on an argument for the existence of God. This twentieth-century philosophy finds its inspiration in Calvin (although the idea can be traced back to Paul) and its expression in the work of Alvin Plantinga, Nicholas Wolterstorff, and William Alston. They contend that belief in God is typically produced, and justifiably so, by something akin to a God-faculty. Alvin Plantinga considers the God-faculty, which he calls the *sensus divinitatis*, as analogous to our other cog-nitive faculties:

> The *sensus divinitatis* is a disposition or set of dispositions to form theistic beliefs in various circumstances, in response to the sorts of conditions or stimuli that trigger the working of this sense of divinity. . . . There are many circumstances, and circumstances of many kinds [glories of nature, grave danger, awareness of divine disapproval, etc.], that call forth or occasion theistic belief. Here the *sensus divinitatis* resembles other

belief-producing faculties or mechanisms. If we wish to think in terms of the overworked functional analogy, we can think of the *sensus divinitatis*, too, as an input-output device: it takes the circumstances mentioned above as input and issues as output theistic beliefs, beliefs about God.[3]

Plantinga thinks that belief in God is neither innate (inborn, or inscribed directly onto the human mind) nor, typically, produced by consideration of arguments. Rather, belief in God is occasioned in various circumstances by the *sensus divinitatis*. For example, when standing on a mountain top, one might simply find oneself, occasioned by one's experience, overwhelmed by the belief that there is a Creator of the world; or feeling guilty, one might have a sense that one is responsible to God.

Suppose we do have a God-faculty, are the beliefs it produces rational?

Cognitive Science

As seen in the previous chapter, cognitive science's investigation into the operations of the mind offers empirical support for Thomas Reid's claim that, in a large number of cases, we have inbuilt cognitive mechanisms, faculties, dispositions, or modules that process information and produce immediate, non-inferential beliefs. The Reidian faculties—perception, the inductive principle and memory, and those that produce beliefs in the external world or other persons and so on—parallel those affirmed by cognitive science. Cognitive science seems to have confirmed that the mind works roughly as Reid conceived it. Cognitive science also seems to support something like Plantinga's *sensus divinitatis*, an innate God-faculty. Or, as the theologically or philosophically unfriendly might put it: what theologians such as Calvin and philosophers such as Reid and Plantinga believed with no evidence whatsoever has found some sort of empirical confirmation! With respect to belief in God, it appears that we do, indeed, have a natural, instinctive religious sense or God-faculty.[4]

The term "God-faculty" is, in the cognitive science of religion, too specific and even misleading. The faculties implicated in religious belief are not a single faculty, nor are they dedicated to producing God-beliefs. The so-called God-faculty is, recall, a conglomeration of faculties that are neither "concerned" with nor aimed at God-beliefs—including, for example, the agency-detecting device (ADD) and the theory of mind (ToM). While these faculties could engender or mediate belief in God as Plantinga conceives, they could also engender beliefs in elves, dwarves, goblins, tree spirits, and witches. Nearly every culture seems to have firm beliefs in spiritual beings and an afterlife. And just as universal human traits such as language and emotion are explained by a mind/brain disposed to language and emotion, it is now widely accepted that recurrent spiritual beliefs indicate that humans are naturally disposed to belief in spiritual beings. Like language, these cognitive dispositions find culturally specific expression, but common to nearly every culture is a firm distinction between the spiritual world and the material world.

The widespread recurrence of religious beliefs, like the widespread occurrence of perceptual and memory beliefs, suggests that the acquisition of religious beliefs is due to the operation of perfectly naturally occurring cognitive faculties. Just as the acquisition of language is natural, so, too, is the acquisition of religious beliefs. Religious beliefs are natural both in the sense that they are easily acquired (typically without reflection or argument) and because they originate with little cultural input. We are naturally disposed to religious beliefs.

Lest one fear that cognitive science (or my presentation of it) offers support for (or is a kind of apologetics for) any particular religion, one should note that, just as we are disposed to acquire a language (but not, say, English), so, too, we are disposed to acquire religious beliefs and practices (but not, say, Christianity). There is no genetic or cognitive determinism here: having a natural disposition to acquire a language leaves a lot to culture (i.e., not our genes) to influence which particular language, and having a natural disposition to acquire religious beliefs/practices leaves a lot to culture (i.e., not our genes) to influence which specific religious beliefs/practices one adopts.

Reid and cognitive science converge: we are cognitively hardwired to believe in agents and other minds (we believe that other people exist, and we form fairly reliable beliefs about other people's thoughts, feelings, and desires). While ADD and ToM may have developed in response to adaptive problems in the Serengeti that suited them to the production of beliefs in animals, mates, and enemies, they have nonetheless also produced beliefs in gods. Sometimes humans have postulated extraordinary agents with extraordinary powers that act for extraordinary reasons through apparent miracles, floods, or thunder. Such big and powerful agents have big reasons for the big things they do. Once one acquires beliefs in extrahuman agents with super-qualities—super-powers and super-knowledge, for example—one has a ready explanation of the causes of super-events.

In short, ADD and ToM produced beliefs in minded supernatural agents that had rich potential for explaining some very important life issues (the weather, for example, or success in war). ADD and ToM are very natural, normal cognitive faculties, which easily engendered, along with the other cognitive faculties mentioned in the preceding chapter, very natural and normal beliefs in gods. Are such naturally occurring beliefs rational?

Reason and Belief in God

If we have a God-faculty, and if the rational stance is correct, then God-beliefs are innocent (rational) until proven guilty (irrational). That is the short of it, anyway. We will defend this claim quickly and then look at the long of it.

There are at least three reasons to believe that it might be proper or rational for a person to accept belief in God as a deliverance of his or her God-faculty, without the need for an argument.

First, and most importantly, *since belief begins with trusting our cognitive faculties (unless or until they are proven unreliable), we should extend this privilege to our God-faculty (unless or until it proves unreliable or one's belief in God is defeated by other beliefs).* If we are, as I have been arguing, permitted to trust our cognitive faculties and the God-faculty is one of those faculties, then we are permitted to trust the God-faculty and accept the beliefs produced by it. So if one's belief in God is produced by one's God-faculty in the appropriate circumstances, then one is rational in accepting that belief unless or until one has adequate reason to cease holding that belief. Before we invest too much in thinking of the God-faculty as a special, perhaps defective cognitive faculty (according to Dawkins, "a built-in irrationality mechanism in the brain"),[5] we need to remind ourselves that the "God-faculty" is nothing more than a fancy name for our perfectly ordinary, very natural, and generally reliable ADD and ToM (along with the other faculties discussed in chapter 2). ADD and ToM are part and parcel of our set of cognitive faculties that typically produce true beliefs. Like our (rational) trust of memory, senses, and testimony, we can and should (rationally) trust the God-faculty and thus its outputs (unless or until we have adequate reason to cease believing them).

Second, *arguments typically play a small role in the lives of most ordinary but perfectly reasonable people.*[6] While philosophers are fond of touting arguments in favor of their philosophical commitments, such as free will or the nature of morality, it is not a requirement of reason that ordinary folks base their views on free will or the nature of morality on philosophical arguments. And while philosophers have developed a good many arguments for free will and the nature of morality, these arguments have not proven compelling. Moreover, equally competent philosophers have developed equally good ("equally uncompelling," one might think) arguments against any position held by other philosophers. But even philosophers hold substantive metaphysical and moral views without science-like support, compelling argument, or even much argument at all (more on that in the appendix). Even supposing that philosophers had compelling arguments for free will and morality, there is a limited number of people with access to or the ability to assess most philosophical arguments. It is hard to imagine, therefore, that the demand for evidence would be a requirement of reason (even for philosophers). Moreover, despite their best intentions and efforts, no philosophers believe in, say, the external world or other persons on the basis of an argument. If belief in God is relevantly analogous to belief in free will, morality, and the external world, then a God-belief delivered by our cognitive faculties needn't require the support of an argument in order to be rational.

The usually impenetrable philosopher Georg Wilhelm Friedrich Hegel charmingly concurs that rational belief in God does not hang on grasping a theistic argument:

> The (now somewhat antiquated) metaphysical proofs of God's existence, for example, have been treated, as if a knowledge of them and a conviction of their truth were the only and essential means of producing a belief and conviction that there is a God. Such a doctrine would find its parallel, if we said that eating was impossible before we acquired a knowledge of the chemical, botanical, and zoological characters of our food; and we must delay digestion till we had finished the study of anatomy and physiology.[7]

Hegel would have celebrated the rationality of my grandmother, a paradigmatic nonphilosophical believer. She would have cackled had I informed her that her belief in God was irrational because she hadn't fully wrestled with the ontological argument for the existence of God. Imagine what her life might have been like if she had acceded to the Enlightenment demand for evidence. Suppose I had hurt myself and gone to her for comfort, and she had said, "I'm sorry, honey, I'm not sure if you are a person, and I only hug or console persons. Until I get a good proof that you are a person, hugs and consolation are out of the question."

One might think that I have unfairly restricted the notion of arguments to, say, sets of propositions that, using a formal logical system, take some members of that set (premises) to properly support another member (the conclusion) of the set. Such a notion of an argument would indeed be unduly restrictive. Arguments can be much more intuitive and implicit or inductive. But I am looking at religious belief from the perspective of the person who does not believe on the basis of an argument, formal or implicit or inductive. This view finds empirical support in contemporary cognitive science and seems not untypical.

Third, *belief in God is more like belief in a person than belief in a scientific theory*. Recall that scientific hypotheses are warranted by way of explanation and prediction. Few of our beliefs, though, are warranted by way of explanation and prediction. I remember drinking coffee a little while ago, I see a tree through my window and note to myself that the sun is shining through, it feels a bit nippy so I get up to put on a sweater, and I see a photo of a recent lunar eclipse. As a result, I form various beliefs: "I drank coffee earlier today," "I see a tree," "I am cold," and "There was a lunar eclipse." These nonscientific beliefs—memory, perceptual, testimonial—are warranted by my properly functioning cognitive faculties. I don't hold any of them because they are the best explanation of my experiences. I don't use them to make predictions. They are immediately occasioned in various circumstances by my properly functioning cognitive faculties. So, too, my belief in God is rational (like my belief in other persons) if it is immediately occasioned in various circumstances by my properly functioning cognitive faculties.

Suppose God is a person rather than a scientific hypothesis—or, more technically, suppose belief in God is more like belief in other persons than belief in scientific hypotheses. The scientific approach—doubt first, consider all of the available evidence, believe later—seems woefully inadequate or inappropriate regarding personal relations. What seems manifestly reasonable for physicists in their laboratories is desperately deficient in human relations. Human relations involve—even demand—trust, commitment, and faith. If belief in God is more like belief in other persons than belief in atoms, then the trust that is appropriate between persons will be appropriate to God. We cannot and should not arbitrarily insist that the scientific method is appropriate for every kind of human practice. The fastidious scientist who cannot turn off the demand for evidence when she leaves her laboratory will find herself cut off from relationships that she could otherwise reasonably maintain—with friends, family, and perhaps even God.

Belief in God, then, is rational simply by virtue of being immediately produced by a set of properly functioning cognitive faculties. We can and

79

should extend to religious belief the respect we accord more mundane but immediately produced beliefs—such as belief in the external world or belief in other minds. We should likewise treat it as we do our other philosophical beliefs. Even if philosophers have role-specific requirements to believe in free will, morality, and God only on the basis of a compelling argument (which I doubt), ordinary folks don't. There is no reason to lay the philosopher's or scientist's insistent demand for evidence, as did W. K. Clifford, on everyone in every circumstance. Belief in God, if it is produced immediately by our cognitive faculties, is, at first glance, rational.[8]

What about the second or even third glances that the rational stance suggests? Is there evidence to support belief in God? Is there evidence against the existence of God? Does the preponderance of evidence, pro and con, confirm or undermine one's initially rational belief in God? Finally, and to return to the theme of the book, do the discoveries in the cognitive science of religion offer reason to cease believing in God or to distrust the God-faculty? If so, despite all of its initial promise, perhaps belief in God is irrational.

The God Delusion

ADD and ToM, natural as they may be, suggest an Aladdin problem for rational belief in God. Recall from the previous chapter the cognitive Aladdin problem—itty-bitty experiential input, massive belief output. We humans are afflicted with a poverty of stimulus: our view of the world vastly exceeds what we can experience or argue to on the basis of that experience. We offered a Reidian conception of rationality as a solution to the cognitive Aladdin problem. It is OK for us to trust those inbuilt cognitive faculties that contribute so substantially to our conception of the world. But we know that sometimes and in certain circumstances those very cognitive faculties fool us—we see oases in the desert where none exist, and we all think our group of people is better than other groups of people (sometimes with horrific consequences). The God-faculty seems especially susceptible to Aladdin problems. When a "face" in a cloud or crops ruined by flood rub the lamp containing ADD and ToM, out pops God!

The titles of many recent books and articles that consider the cognitive and evolutionary psychology of religion suggest that God is an evolutionarily induced figment of our imagination. Richard Dawkins's *The God Delusion* and Daniel Dennett's *Breaking the Spell* are but two prominent titles among many. Among the most forthright are Ludovic Kennedy's *All in the Mind: A*

Farewell to God and Lewis Wolpert's *Six Impossible Things before Breakfast: The Evolutionary Origins of Belief.* Pascal Boyer's *Religion Explained: The Human Instincts That Fashion Gods, Spirits, and Ancestors* conveys a more subtle message; however, given his definition of religion as "an airy nothing," he may have more accurately titled his book *Religion Explained Away.*

What has turned God into a delusion? What has broken the spell of religion? In a few words: cognitive and evolutionary explanations of religious belief. Science has shown, or so it is claimed, that God is a collective illusion or a delusion. So Boyer writes: "In a cultural context where this hugely successful [scientific] way of understanding the world has debunked one supernatural claim after another, there is a strong impulse [in religious believers] to find at least one domain where it would be possible to trump the scientist. But evolution and microbiology crushed all this."[9] Evolution and genetics, it is claimed, have explained God away, crushing rational faith.

The standard evolutionary account of religion—the belief as by-product account—holds that the God-faculty (roughly HADD plus ToM) produces religious belief as a by-product of our cognitive faculties. (While HADD and ADD are equivalent terms, I will typically use the more neutral "ADD" locution.) Religious belief is produced by cognitive faculties that are aimed at other sorts of beliefs—say, the identification of enemies or mates and the determination of their intentions, hostile or friendly. Cognitive science's claim that God is an evolutionary by-product has led some to claim that belief in God is thereby shown to be irrational. Evolutionary explanations of the development of these processes are alleged to show that survival forces, not a supernatural being, cause various religious beliefs and practices. These forces produced agency-detecting devices that were designed to get us to fight or flee when alarmed by a suspicious sight or sound. In short, they were originally designed to get us to behave appropriately when confronted by a possible predator or enemy. If anything should be produced by way of belief, it should be a belief in an animal or human competitor. But when our hypersensitive agency-detecting device (HADD) turns fairly minimal beliefs over to theory of mind (ToM), extravagant and unintended beliefs in spiritual agencies and powers are produced. Spiritual or religious beliefs are the accidental by-product of otherwise effective behavior-producing modules.[10] When applied outside its domain, the God-faculty is, Dennett claims, a "fiction generating contraption."[11] Dawkins concurs: "The irrationality of religion is a byproduct of a particular built-in irrationality mechanism in the brain."[12]

Defeaters

The claim that developments in cognitive and evolutionary psychology prove that belief in God arises from a non-truth-aimed cognitive mechanism or set of mechanisms takes some unpacking. We have a variety of cognitive mechanisms or faculties, some of which are truth-aimed and some of which aren't. For example, we have memory and perceptual faculties that produce beliefs such as "I had a cup of coffee this morning" and "The sky is blue." We also have psychological faculties or tendencies that cause us to believe "My children are smarter, prettier (or, if neither smarter nor prettier, more well-rounded) than other children," or "I am much better-looking and smarter than most people (barring the truth of that, more well-rounded)." The first set of cognitive faculties is truth-aimed while the second is not. But in both sets of cases, the cognitive faculties involved were (very likely) created through evolutionary processes in response to various environmental pressures.

Belief in God, on this view, is like one's belief that one is above average. We have a natural tendency to such beliefs—we naturally think we are superior to other people—but this tendency is not truth-aimed. After all, unlike the residents of Lake Wobegon, we cannot *all* be above average. Perhaps our primitive ancestors needed to feel superior to members of out-groups in order to gain the psychological boost needed to win out in competition for scarce resources with relatively equal out-groups; the key ingredient in successful groups was this psychological boost possessed by our ancestors (which was then passed on to succeeding generations). The discovery of the evolutionary origins of our need to feel superior could undermine the rationality of one's believing that one is indeed superior. So, too, while I might be initially rational with my belief that I am above average, learning that this belief was produced solely by a non-truth-aimed cognitive faculty (and not by my merits) would defeat my rationality.

Philosophers use the term "defeater" for evidence that undermines one's rationality. Consider an example. Suppose I walk into an art gallery featuring (the entirely fictional) Nancy Regine's original anti-drug paintings at a "Just Say 'No'!" exhibition. My initial belief, given the theme, is that Nancy is not a drug user. Suppose, further, that upon entering, I see Nancy and only Nancy in the room and, at the same time, smell marijuana. There are at least two defeaters of my original (rational) belief that Nancy is not a drug user: she is an artist (stereotypically associated with chemically assisted creativity), and she seems to be alone in a room when I smell the marijuana. So while I was initially justified in believing Nancy was not a drug user (based on the anti-drug venue), that initial justification was defeated by the new evidence.

Having acquired evidence that defeats one's initial rationality does not mean that one's original belief could not, ultimately, prove rational. Suppose I now meet Nancy's husband, a judge who is harsh on drug crimes, and that I see, just around the corner, a seedy-looking character with a lit joint in his hand. If I take my undermining set of data (*Nancy is an artist, artists aren't unlikely marijuana smokers,* and *I smell marijuana*) and add to that my new beliefs (*Nancy is married to a judge who is harsh on drug crimes* and *there is a seedy-looking character with a lit joint in his hand*), then my initial belief (*Nancy is not a drug user*) is likely to survive the initially troubling evidence. My seeing the judge and learning about his antipathy to drug users restores my initial confidence that Nancy is not a drug user. Again, if my only evidence were that Nancy is the only artist in a room in which I smell marijuana, then (perhaps) my initial belief, all things being equal, would be diminished. But all things aren't equal, so, upon considering this new evidence, my initial belief in Nancy's non-drug use is restored.

Likewise, one could recover rational belief that one is above average in this or that respect. If you believed that you are an above-average driver simply because of the instigation of your "I am above average" cognitive faculty, then learning that that faculty is not truth-aimed would undermine the rationality of your belief. In order for you to rationally believe that you are an above-average driver, you'd have to acquire some evidence that you are, in fact, an above-average driver (maybe you earn an A+ from the toughest grader at the Above Average Driving School). After learning that the "I am above average" cognitive faculty is not reliably truth-aimed, one's beliefs in one's above-averageness would be defeated unless or until one was able to acquire some independent evidence in support of one's beliefs.

So, too, some claim that we have discovered the ignoble, evolutionary origins of the cognitive faculties that produce belief in God. These faculties aren't truth-aimed, and so the learning of the primal origins of one's God-beliefs undermines or defeats the rationality of belief in God.

By-Product Beliefs

Human beings have wrinkles on the joints of their fingers and circulate red blood throughout their bodies. But neither the wrinkles on our fingers nor the redness of our blood was selected for evolutionarily. Instead, wrinkles and red blood are by-products of traits that were selected for.[13] A by-product trait is not an adaptive trait but a trait that accompanied an adaptive trait. In general, organisms with heritable traits that allow them to produce more offspring (more than those who lack such traits) are likely to see those traits passed on to succeeding generations. Increasingly flexible fingers, the opposable thumb, a pumping heart, and the flow of life-giving nutrients through the blood were certainly adaptive traits. Flexible fingers covered in skin of a certain rigidity, expandability, and thickness were clearly adaptively advantageous. This combination helped early humans grasp fruit, tools, mates, and offspring. When straightened, the skin gathers around our knuckles in a wrinkly way that is a side effect of a finger's bendability and our skin's flexibility. Wrinkles, then, are a by-product of the adaptive attributes related to fingers. The redness of blood likewise has no adaptive function; it is, rather, an artifact of the fact that blood carries hemoglobin (which is adaptive).

Just as there is nothing wrong with wrinkles and red blood, there is nothing per se wrong with by-product beliefs. By-product beliefs aren't, by their very nature, false, irrational, or all in one's mind. For example, the set of beliefs that constitute modern science is, and our moral beliefs may be, by-product beliefs.

Modern science—including, say, the heliocentric view of the universe and atomic theory—is a by-product belief (rather, set of beliefs), a by-product of cognitive faculties that were developed long before, say, 1600. As Noam Chomsky argued: "The experiences that shaped the course of evolution offer no hint of the problems to be faced in the sciences, and the ability to solve these problems could hardly have been a factor in evolution."[14] Modern science runs totally counter to what we see, hear, and feel. We see the sun rise and set, we don't feel the earth rotate, and we don't hear or feel the wind rushing by at thousands of miles per hour (which we might expect if the earth were hurtling through space in orbit around the sun). The chair I am sitting on seems solid, impenetrable, and stable, not a mass of invisible particles buzzing about at super-speeds in mostly empty space. Our cognitive faculties naturally produced folk physics—a geocentric cosmos with ordinary chair-sized objects—not the highly counterintuitive heliocentric and atomic worlds.

Modern science, a very late arrival on the human evolutionary scene, relies heavily on our mathematical abilities and our abilities to think abstractly and causally. Our mathematical abilities arose out of our capacities to count, and abstract thinking arose out of our ability to think counterfactually. Both abilities (but at their most simple levels) were surely adaptive. Consider how counting might have arisen. How many enemies, exactly, are there? How many children do I have? Miscounting in both cases could prove detrimental to one's reproductive success. The ability to count—one, two, three, four—would, when combined with other considerably more abstract ways of thinking, contribute to the discoveries of geometry and the calculus. But these mathematical advances, which again appeared very late in human history, were by-products of simpler cognitive faculties aimed at simpler environmental pressures.

Modern science, with its rejection of geocentrism and its endorsement of quantum reality, runs deeply contrary to common sense. Since we don't feel the earth rotate or hurtle through space, and since a rock feels as though it is one, solid, and impenetrable, the development of modern science required a highly disciplined ability to think counterfactually (or to think counter to ordinary ways of conceiving of the facts). Our ancestors who developed the ability to think counterfactually—to imagine alternatives to the real world and then to mentally play out various consequences—were likely successful in competition with those who were not so adept. "What would happen if we were to attack them from that direction rather than our usual way?" "What do you think would happen if I were to strike these two spark-creating rocks near some very dry leaves?" "What might happen if, instead of throwing rocks with our bare hands, we attached them somehow at the end of a long stick?" The ability to think "what if?" questions such as these enabled increasingly sophisticated actions and planning in anticipation of various possible scenarios in the ancestral environment. Modern scientists availed themselves, in deeply counterintuitive ways, of counterfactual thinking; when combined with the mathematization of reality and a host of other cognitive faculties, such as causal thinking, that did not have modern science in mind, we got (again, very late in human history) heliocentrism and atomic theory and general relativity. Our evolved cognitive capacities—which evolved for picking berries and mates—knew nothing of modern science. Modern science is a by-product belief (an awesome by-product, but a by-product nonetheless).

Morality, some claim, is also a by-product belief. Philosopher Jesse Prinz, for example, contends that our capacity for making moral judgments is "an evolutionary accident."[15] Prinz rejects the view that human morality was adaptive. He argues that if there were an innate moral instinct, humans would have acquired universal moral beliefs. But, he claims, there are no moral universals. He concludes: "Morality . . . is a byproduct of capacities that were evolved for other purposes. . . . There is no mechanism dedicated to the acquisition of moral norms."[16] According to Prinz, we have no innate or native cognitive faculties that are dedicated to morality; morality arose from other, non-morally salient, cognitive capacities such as emotion, imitation, and rule-formation. Morality is the by-product of the alliance of certain feelings—of disgust, say, or shame—with our penchant for making rules. Morality, then, is feelings set to the tune of rules.

Modern science is surely a set of by-product beliefs. Morality may be a set of by-product beliefs. Religion, according to some, is a by-product belief. I discuss these by-product beliefs to raise an important question. Is religion, on the by-product account, rational or irrational? This is partly a function of the particular religious beliefs one has in mind and the extent to which one judges them as relevantly similar to other by-product beliefs, such as science or morality. For example, does belief in God relevantly resemble a scientific theory? If so, does it have the warrant of modern science? If not, does it need to be warranted as scientific beliefs are warranted? Or is belief in God more like our moral judgments? And does it have warrant similar to our moral beliefs? If not, does it have standards of warrant different from moral beliefs? And, in the neighborhood of moral beliefs, how does religious disagreement affect one's warrant for religious belief (in ways similar to our moral judgments)?

Even if religious belief is a by-product belief, that would not by itself show that religious belief is irrational. Cognitive science of religion itself (indeed, all of cognitive science) is a by-product belief—it is not an adaptation, has no evolutionary functions, and was not produced directly via natural selection. CSR and any conclusions one draws about its consequences for the truth or falsity and rationality or irrationality of religious belief are, one and all, by-product beliefs. Let us proceed with caution and without hasty judgment.

Defeating Belief in God

Does uncovering the cause of religious beliefs show that religious beliefs are fanciful expressions of hidden cognitive mechanisms? What if you were to come to believe that you acquired belief in God via processes that involved neither rational reflection nor divine instigation? Instead, you come to believe that it was produced, like beliefs in fairies and elves, by cognitive faculties that are stretched beyond their legitimate domain, beyond their breaking point.

Consider another example of how learning that one's belief was produced by a non-truth-aimed cognitive process undermines one's rationality (while reliance on truth-aimed cognitive faculties supports one's rationality).

Suppose that Dathan throws a birthday party for his brother, Carsten. Just before the party starts, Dathan slips Carsten a Barack Obama Pill, one that produces a vivid visual sensation of Barack Obama (which also renders one incapable of seeing Barack Obama, yet capable of seeing everything else). The visual sensation of Barack Obama is so real-seeming that it is phenomenologically indistinguishable from a veridical perception. It seems as real as any really real visual sensation; but remember, it prevents one from actually seeing Barack Obama (even if he were standing right in front of one). The Barack-Obama-Pill-induced Obama sensation seems every bit as real as the veridical perception I am having right now of my cat sitting in my lap.

Now suppose that Dathan really has invited Barack Obama to Carsten's party. When Barack walks through the door, Carsten exclaims (without, recall, seeing him), "Wow, Barack Obama's at my party. This is the best party ever! Thank you, brother Dathan." Dathan laughs uncontrollably.

Suppose the next day Carsten is told about the Barack Obama Pill. As soon as he learns that his Obama-belief was formed by the Barack Obama Pill (and not by Obama), Carsten's rationality is undermined, removed, defeated. Learning that his belief was caused by neural processes induced by the Barack Obama Pill renders Carsten irrational. Even if Carsten had had a true belief (which he did—Obama had shown up for the party), the process that produced his apparent perceptual belief was not truth-aimed. By learning that his belief in Obama was produced by a non-truth aimed (drug-induced) process, his initial rationality is undermined. Dathan to Carsten: "You just believed Barack Obama was at your party because of the (non-truth-aimed) Barack Obama Pill."

If one were to come to believe that the God-faculty is like the Barack Obama Pill, one's initially rational belief in God would be undermined. There *could* be a god, just as Obama could have been in the room, but belief in God would no longer be a viable intellectual option.

Dawkins to religious believer: "You just believe in God because of a (non-truth-aimed) God-faculty."

Defeating the Defeater

One response to this objection would be to show that God-beliefs, appearances notwithstanding, aren't really like Carsten's Obama-belief. Let us consider very carefully the precise problem with Carsten's Obama-belief and then see if belief in God suffers from this same defect. In order to formulate a response to the claim that CSR defeats rational religious belief, we need a better understanding of what makes Carsten's Obama-belief problematic.

How does Carsten's Obama-belief differ from veridical (that is, truthful or coinciding with reality) perceptual belief? When I genuinely see Barack Obama, my perceptual faculties (vision) convey information to those portions of my brain that process visual information (sensations) and then transfer that information to the portion of my brain involved in believing. Moreover, in veridical perception *I see Barack Obama*—that is, I am in the right sort of relation to the object of my perception (Barack Obama): Barack Obama is the cause of my belief that I see Barack Obama. Carsten had a perceptual belief, but the Barack Obama Pill circumvented the appropriate cognitive processes for the production of perceptual beliefs. Moreover, Carsten was not in the right causal relation to the object of his apparent perception (Barack Obama); the pill, not Barack Obama, was the cause of his belief. Thus informed, Carsten's initial rationality was defeated.

Genuine perception involves both the right cognitive processes (those that can put us in the right sort of contact with their object) and the right sort of contact with that object (better, the object being in the right sort of contact with our cognitive faculties). The cognitive processes involved must bring about the formation of true beliefs about the object *and* bring us into "contact" with the object. My belief that I see Obama must involve perceptual processes that enable my perception of Obama and give me the ability to form true beliefs about Obama, and those processes must put me into perceptual contact with the person that caused the sensation. I cannot rely on hearing or taste to produce the visual sensation required for the belief that I see Obama. Nor can I use reason to produce my belief that I see Obama. My visual faculties must put me into visual contact with Obama. Finally, Barack Obama—outside of my mind, out there, in the world—must be the source and cause of my sensations.

What about other sorts of beliefs? What makes them rational? Similar processes are involved in the production of other beliefs. Let us consider just two more: memory beliefs and person beliefs.[17]

My rational *memory beliefs* must involve my memory faculties, which must put me in the right sort of "contact" with something that actually happened in the past. So my belief that the night of my son Will's birth was dark and stormy must be produced by my memory faculty, and it must be connected in the appropriate way to that dark and stormy night. Suppose that, right after tenderly recounting the circumstances of my son's birth on a stage to a large group of people, I hear fingers snapping and then learn that a hypnotist had inserted that belief into my mind just a few moments before. In those circumstances, this memory belief would no longer be tenable (even though it might still be true).

It is hard to say what exactly we could mean by "the right sort of contact with the past" when considering memory beliefs. We don't come into contact with the past as we come into contact with things in the present. The past is not here, now, in front of us, the way a present physical object is; we cannot touch objects in the past, and we cannot see them. However, if my belief was produced by my memory faculty, and if my memory faculty put me in the right sort of touch with an event in the past, then my memory belief is rational. There must have been some event, something to which I no longer have any direct access, which actually occurred, that is the cause of my memory beliefs. If I am reliably informed that the night of my son's birth was not dark and stormy, my initially rational memory belief would be defeated. Better, if I were to come to learn that my belief that the night of my son's birth was dark and stormy was produced by a non–truth-aimed process (say, a hypnotist), then my initially rational belief would be defeated.

On to person beliefs. Theory of mind (ToM) produces rational beliefs when I come into contact with *personal agents*. ToM does this in two very different ways. First, ToM produces in me the belief that there are other persons. While one might think that perceiving a human body is identical to perceiving a person, it is important to note that what makes someone a person is not her body (things we can see) but her thoughts, feeling, and desires (things we cannot see). If appearing to have a human body were sufficient for personhood, then embodied robots would be persons. But since robots lack thoughts, feelings, and desires, they aren't persons; even with perfect replicas of human bodies, robots aren't persons. So, to make the point, having a human-like body is not the same as being a human person. Being a person requires having or being capable of thoughts, feelings, and desires. But since I cannot see another person's thoughts, feelings, and desires (though I can see another person's body), I cannot see the very thing I would need if I had to have evidence in order to believe in other persons. I can, of course, know that

I am a person because I have direct, introspective access to my own thoughts, feelings, and desires. But I cannot see or feel *your* thoughts, feelings, and desires. And yet, without waiting on evidence (evidence I couldn't possibly attain), ToM immediately produces in me person beliefs whenever I come into contact with another person.

ToM does not just produce the belief that you, for example, are a person. It also produces, without reflection, some beliefs about your beliefs. That is, ToM forms beliefs in me about *your* thoughts, feelings, and desires. I see your face and immediately form the belief that you are happy, I watch your gait and immediately form the belief that you are depressed, or I get a letter and immediately form the belief that you are lonely. ToM is reasonably reliable in generating true beliefs about other people's beliefs. ToM is engaged countless times each day while driving to and from work, while walking at the mall, or while watching television or listening to the radio. I instantly and constantly find myself with person beliefs, engendered by some sort of contact with another person and processed by ToM, as well as beliefs about other persons' beliefs, feelings, and desires (even though I cannot see into their minds).

ToM produces some false beliefs as well. When my kids were young, they would walk into their bedroom at night and get startled by the chalky visage of their great-great-grandfather framed on the wall. His handlebar moustache, sharply focused eyes, and grim, unsmiling visage were simply too much for them to bear. Their tender hearts started beating rapidly; they felt like someone was watching them, ready to jump out of the shadows and sweep them away. They would quickly switch the light on and see, to their relief, that it was just that darn photograph again. As soon as they became aware that they were scared by a lifeless photograph, their belief that there was another person in the room was no longer rational. The problem: the person belief was caused by a photograph, not by a person. ToM produces rational beliefs when they are caused by a person. So if we become aware that the person belief we formed was not caused by a person, it is thereby rendered irrational.

As with memory, determining what constitutes proper causal contact with persons is difficult to conceive. The paradigm case—when I am looking at another human being in perfectly good lighting—is obvious but, again, problematic (for philosophers). Remember, I don't see another person's mind even in the best of light; I just see the person's body. And so I don't *perceive* persons strictly speaking. ToM may be triggered by the perception of a human body in some cases, but it is not a perceptual process (because I don't see the thoughts, feelings, and desires that produce my beliefs in the person's thoughts, feelings, and desires). I see a certain kind of body—a human body—and find myself believing that this is a person (and so treat her in a very different way than I treat, say, kitchen utensils, the weeds in my lawn, or a comfy chair).

While, in the paradigm case, ToM may bring me into the right sort of contact with a physical person who is nearby, contact with persons does not require physical proximity. In fact, I don't even need to see a human body to form a proper belief in or about other persons. I can come into contact with a person through reading a letter that someone wrote to me or an email message that someone sent to me. A young woman can discern the intentions of her beloved by reading skywriting that contains a proposal of marriage. I can tell my wife is angry if cold, canned beets are served on the dinner table or if the door is slammed loudly in another room. I can learn of people from newspaper reports, a biography, or through gossip. I see electron configurations on a television screen that impinge on my retinas, stimulate my rods and cones, and produce the belief that the pope, halfway around the world, passionately cares for the poor. The bottom line remains: ToM works when it produces true beliefs about persons that are caused, ultimately, by a person.

We are now in a position to say what went wrong with Carsten under the influence of the Barack Obama Pill. Carsten's initially rational Obama-belief was defeated when he learned of two things. First, Carsten learned that his perceptual belief was not formed by perceptual faculties at all. Only perceptual faculties produce rational perceptual beliefs (not, say, drug-induced neurochemical processes). Second, Carsten learned that his belief did not involve causal contact with the object of perception—it was not caused by Barack Obama. Thus informed, Carsten's Obama-belief is no longer rational.

And now we are in a position to ask, Are God-beliefs like Carsten's Obama-belief? Are God-beliefs rendered untenable by the cognitive processes that produce the beliefs?

God and Barack Obama

Popular memes assert that God is like a lot of things. God is like Coke: he is the real thing. God is like Allstate Insurance: you are in good hands with God. God is even like Hallmark cards: he cares enough to send his very best. But if God-beliefs are like Carsten's Obama-belief, then coming to learn that they were produced by non-truth-aimed cognitive faculties would undermine one's otherwise rationally held God-beliefs. CSR suggests that very ordinary and some extraordinary experiences or circumstances incited early humans to form beliefs in god(s). Some of these ordinary sorts of circumstances—hearing a thing go bump in the night or a rustling in the grass—can produce agent beliefs when no agents are present. Non-agents like clouds and weather patterns can likewise induce belief in extraordinary agents. The God-faculty seems, then, to pull God-beliefs out of thin air. If God did not factor as a cause in those beliefs and if one is informed of this, then, like Carsten's Obama-belief, belief in God is undermined.

Suppose there is a god with whom some people's beliefs are rightly connected some of the time. Would finding out that on certain occasions the faculties that dispose us to belief in God misfire (producing belief in God without God being the cause of the belief) undermine rational belief in God?

If one were to come to believe that God was not causally involved in the production of one's own belief in God, then I think one could no longer rationally believe in God. Moreover, if one came to believe that one's God-beliefs were the result of non-truth-aimed cognitive processes, then I likewise think one could no longer rationally believe in God. But has CSR shown that God was not causally (even if not immediately) involved? In order to show that God-beliefs *are* like Carsten's Obama-belief, one would have to show that God was not causally involved in the production of the beliefs.

The issue of how God might properly *cause* our God-beliefs is, as one might expect, complicated. Recall that it is hard to say how the past might cause memory beliefs, and it is even harder to say how persons cause person beliefs. We can come into contact with a person through letter, email, television, radio, internet, smoke signals, and many more ways; but remember, we never come into direct contact with minds—the peculiar aspect of persons that makes them persons. So, again, it is hard to say precisely how persons and the past cause our beliefs about persons and the past. But an actual event in the past must be the ultimate cause of my memory belief, and a person must be the ultimate cause of my beliefs about persons.

How might God be the cause of God-beliefs? God might cause a disposition in us to believe in him, one developed through evolutionary processes and discovered by cognitive science of religion. One could have a God-caused religious experience, an experience that one finds difficult to put into words. Someone who had such an experience could also tell people about it; warrant for their belief would transfer through testimony. Reliable chains of testimony can go back for many, many years and through many, many people as long as they began with someone who did have a God-caused religious experience (and the testimony did not get distorted by others). God might send us some sort of revelation (say, a letter communicated through human authors) or create in us a conscience that connects us with the ultimate source of goodness. If God is the author of nature or our greatest good, God could use our teleological faculties (recall our disposition to see purpose in many things) or our deepest desires to move us to himself. In short, there are countless ways in which God might properly be the cause of one's rational belief in God.

While God himself may not have been the *immediate* cause of God-beliefs, God may nonetheless be the *ultimate* cause of those beliefs. If God is the first and originating cause of the universe (including all natural laws), and if God were to guide or direct the evolutionary processes so that they produced a God-faculty so that people could and would come to form true beliefs about God and come into an appropriate relationship with God, then God would be the ultimate cause of our God-beliefs. And so our God-beliefs would be caused by their proper object—God. God may not be directly or immediately involved in the production of God-beliefs, to be sure. But the cause of one's beliefs need not be direct or immediate. If God is the ultimate cause of true beliefs about God, God-beliefs can be rational—even if they are produced by natural processes and God is not in the immediate neighborhood. Learning that the immediate cause of God-beliefs involves natural faculties would not show that our God-beliefs were untenable after all. In order to show that, one would have to show that God was not the ultimate cause of our God-beliefs. And that has not been done.

If there is a God, then becoming aware of the natural processes that produce belief in God does not show belief in God to be irrational. God-beliefs, therefore, may not be like Carsten's Obama-beliefs. If God-beliefs are produced by cognitive faculties that are adequate to their object, and if God-beliefs are ultimately caused by God, then one can rationally believe in God. If there is a God, learning of the natural processes that produce those beliefs does not render God-beliefs irrational.

In order to know or reasonably believe that God is not the ultimate cause of one's God-beliefs, one would need to know or reasonably believe that God does not exist (not just show that natural processes are involved in the production of God-beliefs). But Dawkins and Dennett and others have not done that. They have done nothing more than show that natural processes are involved in the production of God-beliefs. But that is not sufficient to undermine the rationality of God-beliefs; to show that, they would need to show someone a compelling argument against the existence of God. And they have not. Since they don't believe that God exists, they likewise don't believe that God is the cause of God-beliefs. But they have not shown that he is not, and they cannot do that simply by pointing out that God-beliefs involve natural processes. It is no small task, then, to show that religious beliefs are irrational.

Consider an analogous case against belief in others' minds. Suppose a cognitive scientist, let us call him Faniel Fennett, argued that cognitive, neuroscientific, and evolutionary accounts of mind-beliefs show that belief in others' minds and personhood are the result of natural selection because of their adaptive utility. Moreover, Professor Fennett argues that we have special cognitive faculties that, under normal circumstances, give us such beliefs, and that we even know a lot about the brain structures that undergird these faculties. Professor Fennett concludes that we now have a defeater for belief in others' minds and so to continue to believe in them is irrational.

But if we have a cognitive faculty, say, theory of mind (ToM), which produces true beliefs about other people's thoughts, feelings, and desires—beliefs caused by other persons and produced by the proper cognitive process—then learning that ToM was naturally produced does little to defeat the warrant of those beliefs.

We cannot know if ToM is defective unless we already know that other people lack minds. Likewise we cannot know if the God-faculty is defective unless we already know that there is no God. If one does not believe in God, one will believe that God is not the ultimate cause of God-beliefs. But this piece of autobiography tells us more about the atheist's personal beliefs than about the logic of the situation. And the atheist's personal beliefs on this matter scarcely constitute evidence that there is no God. Unless or until the religious believer has sufficient evidence that God does not exist, then she has no reason to believe that God is not the (ultimate) cause of her belief. (The non-theist might argue that the natural explanation without God is better than the natural explanation with God, because the former is simpler [or might argue that adding God as ultimate cause is ad hoc or superfluous].)

For the theist, however, learning that hers and others' beliefs involved very natural cognitive faculties (even ones developed evolutionarily and accidentally) is not sufficient to defeat her rationality.

Religious Experience

One's experience of God could continue to ground one's rational belief in God even if one were to become aware of the natural processes involved in the production and sustenance of one's belief in God. Moreover, disagreement of even very smart people on this matter needn't undermine one's rational belief in God. Let us consider a couple of examples.

Suppose you read that wild turkeys were long ago driven out of the state of Michigan. Every book that you read gives good reason for believing that the wild turkey has disappeared from the state. On the propositional evidence that you have acquired through books by relevant authorities, it is reasonable to believe that wild turkeys no longer exist in the state of Michigan. But suppose you wake up early, walk out into your backyard in Michigan, and come face-to-face with a flock of wild turkeys. At that moment, you have good experiential reason to believe that wild turkeys live in Michigan. Your reason is not propositional; it is experiential (you see a turkey). Your reason is not an argument (unless you could turn "What I see, I see" into an argument). You simply see a wild turkey and find yourself believing that there is a wild turkey before you. Your belief is reasonably and independently grounded in your visual experience, not in a propositional argument. Wild turkeys activate your cognitive faculties in such a way as to immediately and noninferentially produce belief in the existence of wild turkeys. While the expert writers of the books and articles on wild turkeys may disagree with you, you can rightly say "So what?" They didn't see what you saw. They disagree, but both of you are rational (until the experts visit and see your turkeys themselves).

Suppose, to move us closer to the God case, your mother tells you and your brother and sister that your father went down with his ship at sea; she also shows all of you a news article that reports your father's demise. A few years later, you are walking on a beach in Panama and see your father, the proud beneficiary of an insurance scam. You walk up to him and talk with him. While your brother and sister are at home, rationally believing that their father is no longer alive, your experience of seeing and talking with your father grounds your belief that your father is indeed alive. You don't have an argument that your father is alive, and you don't need one. You might not be able to persuade your brother and sister, but so what? Your inability to persuade them does not preclude your rational belief that your father is alive. So you respectfully and reasonably agree to disagree.

In short, if one has an experience of God or if God is the ultimate cause of one's God-belief, then one can remain rational even if that belief is mediated by a natural process and even if really smart people (who have not had the same experience) disagree.

Simplicity?

The philosophically reflective non-theist might argue that the principle of simplicity requires one to rationally reject any supernatural involvement in God-beliefs. She might hold that if a natural explanation of religious belief explains religious belief, one can no longer rationally believe in a supernatural explanation. In short, according to the philosophically reflective non-theist, a fully successful cognitive and evolutionary psychology of religion undermines rational belief in God.

We aren't at that point, of course. We have some ideas about how religious beliefs and practices *may* have formed through very natural cognitive processes, and we have some ideas about how the mind admits and shapes God-beliefs. But beyond some very highly educated guesses, we simply don't know the origins, natural or not, of the earliest religious beliefs. Moreover, as we will see in later chapters, we don't know of any particular person the origins, natural or otherwise, of her beliefs. We don't have much access to the earliest human history, and we don't have much access to another person's psyche. And so we don't have much sense of how religion was acquired in general or in particular.

But suppose some future science were to develop a fully adequate and universally accepted natural explanation of religious beliefs. Would simplicity require one to give up, on pain of irrationality, one's religious beliefs?

Simplicity is valued in scientific theorizing to prevent needless complications and explanations. Mathematically simple and elegant theories are preferable to more complex theories. But more to the point for this discussion, once a particular set of data is adequately explained by various theoretical entities, one should not (because one need not) postulate any additional entities whatsoever. For example, if quantum phenomena can be fully and adequately explained by atoms, then don't go around looking for anything extra to explain quantum phenomena; there is no need to populate the world with extraneous or superfluous unseen particles—unless, of course, there are additional data that require us to dig deeper into reality for other sorts of entities in order to explain the new data.

Physicists were forced by new data to postulate, in addition to atoms, their constituents—protons, neutrons, and electrons (and, later, even more subatomic particles such as quarks). But scientists should not postulate or accept any additional entities unless they are required to by the data. So, to cite Occam's razor, scientists should not multiply explanations beyond necessity.

With respect to the God-faculty, then, one might argue that if there is a fully natural explanation of religious belief, then it is explained. Full stop. While one *can* put a theological overlay on the natural processes that produce belief, one should not bring in the supernatural unless it is rationally required; one *should not* because one *need not* bring in the supernatural.

There is no reason, as far as I can see, to appeal to a god to explain any data in the cognitive science of religion. The scientific practice of the cognitive psychology of religion, following Occam's razor, should not countenance the existence of God in their scientific theories concerning the God-faculty. Agreed.

While appeals to the supernatural aren't *scientifically* necessary to explain God-beliefs, the question posed by this book is not about how science is best practiced or what one should believe as a scientist. Scientists when doing science should invoke and abide by the principle of simplicity, and so appeals to the supernatural *are out of place* in science. One is simply not doing science unless one restricts the range of permissible explanations in this way. Simplicity, along with a host of other values such as predictive power and fertility, is a methodological commitment of science.

Science is also, by its nature, methodologically natural. That is, it looks for the natural processes that are operative within nature, eschewing anything that smacks of the supernatural. And rightly so: in setting aside supernatural entities and forces in its inquiries into the natural world, science has achieved a remarkably deep understanding of the natural processes involved in nature. But by setting the supernatural aside at the beginning, it cannot reasonably assert that, in the end, it has disproved the supernatural.

Suppose a group of brilliant mathematicians had decided to leave out the odd numbers in all their calculations. And suppose, after years and years of remarkably fruitful mathematical advancements, one of the mathematicians informed a group of young and impressionable undergraduates that mathematics had demolished the odd numbers. Such a pronouncement would and should be treated with guffaws. If, as a matter of mathematical methodology, you leave the odd numbers out, you aren't going to end up with odd numbers.

If one assumes at the outset that there is no God, then it should come as no surprise when one's theories fail to include gods.

Finally, for most people, God is not a hypothesis that provides a better or more complete scientific explanation of religious beliefs; God, for most people, is not a scientific hypothesis at all. As with most nonscientific theorizing, the principle of simplicity does not apply. While the principle of simplicity is useful both inside and outside the domain of science, it should not rule our believing life.

Suppose we were required by reason to follow the principle of simplicity in all areas of human inquiry. If so, I should no longer believe that any other persons exist. I can fully explain the data of other persons by believing that they are simply creations of my mind (without believing in their existence independent of my mind). I may see your face, notice your frown, and hear the plaintive whine of your voice. But the simplest hypothesis is that only I exist and that you and other "people" are simply figments of my imagination. If I can explain my person beliefs by belief in just one person—me—then simplicity requires that I not postulate the existence of other entities (like you). With respect to other persons, I don't seek the simplest explanation; I believe what seems true (even if it complicates my metaphysical picture of the world).

Simplicity is not the only virtue to consider when developing a rational belief system.

Take our beliefs about the external world. There is no need to explain my beliefs about the world by postulating an enduring physical world outside my mind. The simplest view is that only I exist and my sensations of colors and textures and sounds are nothing more than my sensations (mental events). I don't need to go beyond mental events to a more complicated postulation of the material causes of those events. I don't need to go beyond my tree-sensations to a tree.

Such is the stuff of simplicity: only I exist—no other persons and no external world are necessary. But, once again, I am not seeking the simplest theory to explain my sensations; I simply believe what seems to me to be true. I believe in other persons and the external world.

However, if I were to take other persons and the external world as quasi-scientific hypotheses offered to explain the data of my sensations of persons and the world, the principle of simplicity would preclude their rational acceptance. If I can account for the relevant experiences without appeal to anything but myself, and if I should not multiply entities beyond necessity, I should believe that only I exist.

I might go even further. After all, what am I but a conscious self that persists through time (and through a host of bodily changes). But it would be simpler to believe that only my sensations exist, not a conscious, enduring self; selves, according to some philosophers and scientists, are convenient fictions that don't correspond to anything in reality. If it is simpler, numerically speaking, to think there is only one person with lots of sensations, it is even simpler to think there are zero persons and just sensations.

But I don't conduct my belief in other persons or my enduring self according to the scientific method. As a general belief policy, nothing seems sillier. So I don't.

I don't take other persons, myself, or the external world as *hypotheses* that explain some set of data. And I don't accept other person-beliefs or external world–beliefs on the basis of hypothetical reasoning with appeals to simplicity. In fact, I don't reason to them at all. I take them to be true (even though I cannot prove them and they aren't the simplest hypotheses). Of course, even scientists assume other persons, their own self, and the external world—even though they aren't the simplest hypotheses that adequately explain the data.

Let me offer one more example. I believe that it is wrong to murder, steal, and break promises and that it is right to be generous, show compassion, and encourage human flourishing. Some scholars hold that our moral beliefs are a trick played on us by our genes, that evolutionary forces outfitted us with moral beliefs because groups with those beliefs were more successful (in terms of reproduction and survival) than groups without those beliefs. Groups with such pro-social beliefs were more cooperative than groups without; as such, groups with moral beliefs gained cooperative benefits—shared childrearing, working together to raise and store crops, division of labor, and success in war. The more cooperative one's group, the more likely one would be to get adequate food, shelter, clothing, and protection from enemies. Since morality conduces to cooperation, morality is evolutionarily beneficial. Should I, therefore, think of morality as a mere fiction?

In cognitively responding to the various experiential pushes and pulls of reality and to life as it presents itself to me, I am doing my best to understand what's in the world outside myself. I come to my understanding of the world through my cognitive faculties, assuming that what they deliver to me is the sober truth, unless and until I have reason to reject their deliverances. I believe that other people and an external world and a distant past are presented to me through my cognitive faculties, even though I have no good arguments for them and even though my meta-physical worldview would be simpler without them. In my attempt to understand the world in all its richness, I'm not seeking the simplest explanation of my experience. I'm a person doing my best to grasp the world through my admittedly finite and fallible human constitution, not a scientist in a laboratory developing the simplest comprehensive expla-nation of a body of data.

For such a believer, simplicity would be just as irrelevant in judg-ments about God as is it in judgments about other persons or the external world.

One Final Caution

Does learning of the cognitive science of religion defeat one's rational re-ligious belief? If one were to come to believe that one's belief in God was nothing but the product of one's misfiring cognitive faculties, then one should, on pain of irrationality, give up one's belief in God. But CSR has not and could not show anything like that at all. It can show what cognitive faculties are involved in thinking about God, and it can suggest how reli-gious beliefs *might* have arisen in some of the earliest human communities. But it has not, indeed cannot, show that God is not the ultimate cause of one's belief in God.

Moreover, at least at this stage, the cognitive science of religion knows little of the origin and sustenance of the religious beliefs of, say, the ancient Hebrews, Paul, Muhammad, Confucius, or my grandmother. While we might know that their beliefs were mediated by ADD and ToM, we know little beyond that. We have no idea if the origin of their beliefs was the mis-firing of ADD and ToM, reflection on the nature of the cosmos, or an en-counter with the divine. If one is an atheist, one will reject the latter option out of hand. So be it. But the atheist's personal beliefs are irrelevant to the rationality of the religious believer.

We think God with our brains (minds). A science that identifies the parts of our brains involved in God-beliefs should not come as a surprise. Some parts of our brains may on occasion produce, without divine instiga-tion, God-beliefs. But learning of the occasional fallibility of an otherwise reliable cognitive faculty does not defeat one's rationality in holding beliefs produced by that faculty. My faculties may sometimes produce perceptual hallucinations, but I needn't reject the deliverances of my perceptual faculties. One cannot know if ADD and ToM are always misfiring with respect to God-beliefs unless one already knows that there is no God. In order to undermine one's rationality, one would have to show that the cognitive faculties involved are misfiring *and* that there is no God.

CHAPTER 5

Against Naturalism

Is It Really All Good?

It has often been acknowledged that the person with an experience is never at the mercy of the person with an argument. While some might come to believe in God initially by means of philosophical arguments, most don't. For most, the *experience* of God's existence, no matter how unremarkable or limited that experience, often precedes, or at least coincides with, philosophical argument. And of even greater significance is the fact that, without the experience, intellectual acknowledgment is often sterile. It is reasonable to believe that a God driven by love to create humans with the capacity to know and love him in return would not be content with the mere intellectual acknowledgment of his existence.

And yet, as truth-seekers, believers want to know if what they feel is real. Believers want to know if they have touched God and not just their cerebral cortex. After the challenges of cognitive and evolutionary science, believers want to know if God is more than Santa Claus.

Believers might come to understand, in the face of these challenges, that it is *possible* that they could be rational in their belief in God (assuming that God exists and that various other conditions have been met, blah, blah, blah), but believers sometimes want more. Is there any reason to think that God really exists? Or is God nothing but a brain spasm, bequeathed to us by our primitive, helpless, and unreflective ancestors?

Belief begins with trust: we are rationally permitted to accept beliefs given to us by our cognitive faculties; we get that. We have a God-faculty, and we can trust it and the beliefs it gives us unless or until we have good evidence to reject it; we get that, too. But the specter of explaining God away looms large and close; God seemed so much more vibrant when science required divine assistance. Even if we cease believing in God as a scientific hypothesis in competition with increasingly better naturalistic (non-god) scientific hypotheses, God, like the grin of the Cheshire cat, seems to be ever so slowly disappearing.

As truth-seekers, we should be willing to revise our beliefs in the face of counterevidence. Yet, I've argued so far, it is possible for some people who believe in God to maintain the confidence of their initial commitments in spite of alleged challenges to that belief from the cognitive and evolutionary psychology of religion. As some of my younger friends say, "It's all good, it's all good."

But those friends typically say, "It's all good," when it's not all good; in fact, they seem to say that only when it's pretty bad. What they really mean is "It's not all good (and it's maybe even bad), but I'll put on a good face and persevere." Some religious believers, after grappling with philosophers and scientists who have attempted to explain God away, may be unbowed but not unbloodied. And while they may say, "It's all good," they may feel inside like it is not, all the while hoping that there is more to be said in favor of belief in God. They may not need intellectual assurance to be rational, but they sure want some. They want more than just the possibility of being rational—they want some assurance that they have hitched their faith-wagon to the star of truth.

As truth-seekers, we want supporting evidence in favor of our most cherished beliefs. And we may want it more now, with all these brilliant scientists seeking to explain gods away, than we ever have before. Again, we may not need it in order for our belief in God to be rational. But, dammit, we want it. Is there any reason to think our belief in God is true? Is there evidence that favors religious belief? If one believes in God because one's God-faculty was triggered in the appropriate circumstances—on the basis of a religious experience, a feeling of guilt and subsequent feeling of forgiveness, hearing a good sermon, or simply an overwhelming sense that God created all this—and it has survived the challenges of cognitive science, there may also be evidence available to strengthen one's initial belief. Such evidence would be a sign, a confirmation that the belief is true.

In this chapter, then, I consider how one might logically bolster one's already-held belief in God through, of all things, evolution. We have seen in previous chapters that rational belief in God should not be perceived as contradicting the theory of evolution. But even if we grant this, we might still think that, when it comes to comparing the relative merits of theism and naturalism in light of evolution, the best that theism can hope for is a tie or a wash. In the eyes of many, naturalism seems to be a natural counterpart to the theory of evolution. If evolution made naturalism intellectually respectable, how could evolution be used to argue against naturalism?

Confirming Evidence

No philosopher has ever developed a deductively valid or inductively strong argument in support of the beliefs that other minds exist, that the external world exists, or that the future will be like the past (the belief output vastly exceeds the informational or evidential input). If we were to rely on reasoning to rationally justify our belief that other minds exist, that the external world exists, or that the future will be like the past, we should be skeptics about all of these beliefs. Fortunately for us, however, we have been supplied with cognitive faculties that justifiably and immediately produce belief in other minds, the external world, and the future being like the past.

However, for one who believes in the existence of other minds through their theory of mind, every gesture, word, and deed of another person would confirm one's original belief. While person-like behavior does not constitute a compelling argument that there are other minds, it does confirm one's already-held belief in other minds. And while tapping on a table or getting pricked by a pin could not prove the existence of the external world, such mundane experiences do confirm one's already deeply held belief in the external world. And when a scientist assumes that the future will be like the past (the principle of induction), every successful prediction confirms that original commitment (which none could prove). In short, evidence might confirm a belief that one could never prove in the first place.

We can learn the dynamics of belief confirmation from a common cop show. Suppose a cop has some initial sense of the guilt of a perpetrator based on, say, eyewitness testimony (being picked out of a lineup, for example). Given the unreliability of eyewitnesses, that initial sense of guilt is nowhere near certain. Each piece of accumulating evidence, however, confirms the guilt of the perp—fingerprints, DNA, additional witnesses, motive, epithelials, and various other kinds of forensic evidence. As each piece of evidence comes in, the cop's initial conviction about the perp's guilt is incrementally increased. If this cascade of incriminating evidence induces a confession, the cop's conviction of guilt becomes almost a certainty.

Confirming evidence both psychologically strengthens and evidentially increases one's confidence in one's original belief. While psychologists (rightly) remind us of the temptation to seek and too easily find "evidence" in favor of our cherished beliefs (the creaking sound in the attic "confirming" belief in ghosts; the tepid comment of one's teacher "confirming" belief in one's own genius), there are objective methods for assessing the strength of evidence in support of a belief. So, in order to avoid the sort of confirmation bias that simply rationalizes a silly belief on the flimsiest of grounds, we will rely on objective evidence and philosophically sound methods of reasoning—all the while conceding that confirmation falls well short of conclusive proof.

Is there evidence that might confirm and strengthen one's already rational but not certain belief in God? In support of belief confirmation, and in the neighborhood of topics canvassed in this book, let us consider the evolutionary argument against naturalism.

Explanation and Expectation

Before turning to the argument itself, we need to first consider how to understand confirming evidence. How are we to judge among competing hypotheses, especially among the god or no-god hypotheses? While there are several methods of weighing different hypotheses given a body of evidence, we will use an intuitively plausible and widely defended method called the *expectation principle*.

Let us proceed by way of example.

If you were out walking and noticed someone with long hair and dirty jeans with a Grateful Dead patch sewn on a pocket listening to Grateful Dead music from a boombox resting on their shoulder and you thought to yourself, "There goes a fan of the Grateful Dead," you were using the expectation principle.

Let us look at the expectation principle just a bit more formally. According to the expectation principle, a set of data (D) favors one hypothesis (H_1) over another hypothesis (H_2) under the following conditions:

If H_1 were true, one should expect D to be true.
If H_2 were true, one shouldn't expect D to be true.

The expectation principle asks, "Under which hypothesis should one more *expect* the data to be true?" A good explanation of some set of data is one that would lead you to expect the data to occur. In the first example, you quickly affirmed the fan-of-the-Dead hypothesis as the one that would lead you to expect the data (and you probably didn't need to consider any other alternatives).

There is more to confirmation than just the expectation principle. Consider another example. You stop by your friend Jan's house to drop off a

CD she lent you. A dozen cans of yellow paint are in Jan's yard. A ladder is propped against the side of her house. Let us take as data:

> D: A dozen cans of paint are in Jan's yard, and a ladder is propped against the side of her house.

Given this body of evidence, D, let us examine two competing hypotheses:

> H_1: Jan is painting her house.
> H_2: Jan is not painting her house.

Under which hypothesis would we most expect D? H_1, of course. If Jan were painting her house, we would expect to see cans of paint and a ladder. If Jan were not painting her house (H_2), we certainly would not expect to see a ladder propped up against it and twelve cans of paint. So, given the data, we have reason to accept H_1 rather than H_2, or, slightly more formally, the data confirm H_1 relative to H_2.

But is there not a whole host of hypotheses available, each of which would equally well lead us to expect the data? While the expectation principle would surely eliminate some hypotheses—those that would not lead us to expect the data—there are countless hypotheses that would lead us to expect the data. How are we supposed to decide among those competitors?

This leads us to an important desideratum when evaluating competing hypotheses: the hypotheses under consideration must also have *some likelihood of being true independent of the data.* Consider a third hypothesis:

> H_3: Jan is constructing a tower of paint cans to ward off an invading horde of yellow-fearing aliens.

H_3 *would* lead us to expect the data. So does the data confirm H_3? H_3 is not confirmed because H_3 is not a viable hypothesis; its lack of viability prevents it (and the countless nonsense hypotheses like it) from getting onto the table of rational consideration. Unless we had previously learned that Jan had an irrational fear of aliens, H_3 has no likelihood of being true independent of the data. We routinely judge the relative initial plausibility of hypotheses against our general background knowledge—our basic beliefs about how things work in the universe (usually without even knowing it). So while ghosts and goblins could explain the things that go bump in the night, they fail the antecedent likelihood test because they don't match up with our notion of reality. Consideration of the independent likelihood of a hypothesis eliminates most from consideration.

One of two competing hypotheses will be confirmed by the data, assuming that both have some initial plausibility that makes us take them seriously as candidates, if it would lead us to expect the relevant data (and the other hypothesis would not). A belief that survived those two tests would come out the stronger for its examination.

Thus lightly armed with the expectation principle, let us turn to the evolutionary argument against naturalism and see if it might confirm one's initial belief in God.

Evolution and Naturalism

What the evolutionary argument against naturalism attempts to show is that a naturalist who accepts the theory of evolution has reason to doubt the reliability of her cognitive faculties and, consequently, the veracity of her beliefs. The problem, as it turns out, is not the theory of evolution, but naturalism. But our commitment to the reliability of our cognitive faculties and the truth of many of our beliefs (including belief in the theory of evolution) should be stronger than our commitment to the truth of any philosophical view, including naturalism. If evolution plus naturalism entails that we should be skeptics about most of our beliefs, then it is naturalism that we should discard, not evolution.

A central claim of naturalism is the denial of any purpose to the cosmos. It follows automatically from this central claim that our cognitive faculties don't have the purpose of giving us true beliefs, since they have no purpose at all. Naturalism's rejection of purpose means that unguided evolution is indifferent to true beliefs. The only things that unguided evolution "cares" about are traits and behavior that are conducive to survival. For the naturalist, evolution has no goals in mind and does not design anything. Instead, some creatures survive and reproduce better than others. And the behavior and traits that allow them to do so are then passed on to the next generation. And, as Stephen Stich writes, "Natural selection does not care about truth, it only cares about reproductive success. And from the point of view of reproductive success, it is often better to be safe (and wrong) than sorry."[1]

The moose didn't develop antlers *so that* it has a handy weapon, and the bat didn't develop wings *so that* it could be the only flying mammal.

But developing antlers enabled moose to injure or kill enemies and so live longer, and the ability to fly enabled bats to reach higher food sources and so live longer (longer, at least, than their no-antler moose ancestors and no-winged bat ancestors, thus increasing their reproductive success). There is no "*so that*" according to naturalism—no purpose, only blind chance. As Dawkins writes, "Natural selection, the blind, unconscious, automatic process which Darwin discovered, and which we now know is the explanation for the existence and apparently purposeful form of all life, has no purpose in mind. It has no mind and no mind's eye. It does not plan for the future. It has no vision, no foresight, no sight at all."[2] For the naturalist, there is no "*so that*" in nature; things just happen. Without a God or cosmic purpose, our cognitive systems couldn't have a purpose at all, let alone the purpose of providing us with true beliefs.

The upshot: if our cognitive faculties were the product of unguided evolution, then their evolutionary function is survival. These faculties did not develop *so that* we could have true beliefs. Rather, the most likely explanation is that we developed our cognitive faculties because they improved our ancestors' reproductive fitness. Given this, however, we have good reason to doubt that their function would be to produce true beliefs.

Darwin himself expressed similar worries regarding the reliability of our faculties. In a famous letter he confided to a friend: "The horrid doubt always arises whether the convictions of man's mind, which has been developed from the mind of the lower animals, are of any value or at all trustworthy. Would anyone trust in the convictions of a monkey's mind, if there are any convictions in such a mind?"[3] Given apparently blind evolutionary processes, even Darwin worried about the human ability to grasp the truth.

We might initially be inclined to think that the opposite is true and that evolution would select for true beliefs. After all, how could knowing truths hurt a creature's likelihood of surviving? If anything, it might seem that knowing more truths would only provide a reproductive advantage. If, for example, I am a hunter-gatherer, looking for my lunch, it seems plausible to assume that knowing the truths about various matters, such as the number of predators that are chasing me or which plants are poisonous, will give me an advantage over my fellow prehistoric humans. Perhaps then we should follow Willard Van Orman Quine, who held that "creatures inveterately wrong in their inductions have a pathetic but praiseworthy tendency to die before reproducing their kind."[4] In laymen's terms: people with false beliefs tend to die off before they have kids.

Is it not reasonable to assume, with Quine, that evolution selected for true beliefs? We are here, we have survived, and we believe lots of truths (truths that our species relied on to get us to this point). So we must have evolved truth-sensitive cognitive faculties. Is it not reasonable to think that our best bulwark against survival pressures was true beliefs? Wouldn't we, then, expect evolution to produce in us cognitive faculties that conduce to the truth?

We have Quine, on the one hand, who claims that evolution would have shaped us with truth-conducive cognitive faculties, and, on the other hand, Darwin, who worries that evolution is infertile soil for human knowledge (and, as we'll see in the next sections, Richard Rorty, Friedrich Nietzsche, Patricia Churchland, and Michael Ghiselin). Of course, we could find others on Quine's side, and philosophy is not a democracy. Can we do a better job of affirming Darwin's worry and rejecting Quine's bold assertion?

One way of warming up to these issues would be to remind ourselves that we do indeed have some adaptive but false beliefs. For example, studies have shown that men overestimate how interested women are in them (and women's interest in having sex with them), whereas women overestimate a man's interest in staying around and caring for a baby (after they have sex).[5] It is not hard to understand the adaptive advantages of the different beliefs.

Let me offer a less flip example: our attribution of color properties to objects is, if contemporary theories of vision are correct, systematically false yet remarkably useful. Consider the redness of a rose. The physics of color holds that the rose (in particular, its electrons) absorbs most of the light waves that hit it and reflects only the unabsorbed red portion of the light spectrum; we, in turn, project the color red (which is not a property of the rose) onto the rose. We have evolved the ability to register and then project various wavelengths of light onto objects like roses; strictly speaking, though, the rose is not red. Surely our ability to discriminate color as we do aided in our ability to fight, feed, flee, and reproduce. But our color attribution is literally false (but eminently useful). We could make a similar case for sounds and tastes.

So it is possible, and on a wide scale, to have false but adaptively advantageous beliefs. But being mistaken in some cases, even on a wide scale, is not tantamount to thinking we could be mistaken about all or almost all of our beliefs. These few sorts of examples should not undermine our confidence in our other cognitive faculties.

So, for a defense of Darwin's worry, we must look deeper. Is there any reason—reason beyond these few examples—to think we could have been *systemically* deluded? Or perhaps this is a better way to put it: we have reason to question some of our evolutionarily induced cognitive faculties, but is it possible for evolution to undermine all confidence whatsoever we might have in our cognitive faculties?

Before we can address this issue directly, we need to muddy the philosophical waters a bit more. The problem with evolution and knowledge may be more complicated than it first appears. One's understanding of evolution and its cognitive consequences shifts on a dime depending on the worldview within which one embeds it. According to Alvin Plantinga, the knowledge problem we have been discussing is not with evolution itself but with evolution embedded within a naturalistic worldview; naturalistic evolution, Plantinga argues, constitutes a problem for knowledge.

What, then, is naturalism?

Metaphysical naturalism holds that there is no such person as God or anything like him. According to naturalism, nothing exists but spacetime, and material objects and events in spacetime. Because there is nothing beyond nature, beyond spacetime, there is no supernature. In this way, naturalism entails that there is no ultimate purpose or design in nature because it denies any Purposer or Designer.

For the naturalist, as noted above, evolution precludes purpose. Some creatures simply survive and reproduce better than others, and the traits and behaviors that help them survive and reproduce may then be passed on to succeeding generations. End of story. None of this happened for a purpose.

For example, to reiterate a point made earlier, the duck did not develop webbed feet so that it could swim, and the giraffe did not develop a long neck so that it could reach leaves higher up in trees. Granted: developing webs enabled ducks to swim faster and so to live longer, and increasing neck length enabled giraffes to reach higher to gather more food and so to live longer; those better-webbed ducks and longer-necked giraffes were better able to reproduce than their less adept peers and to pass on their genes to succeeding generations. But those things didn't happen *so that* they could pass on their genes. As noted above, there is no "*so that*" according to naturalism; there is, according to naturalism, no purpose. Naturalism does not do things *so that*; things just happen for no reason.

If naturalism is true, then humans did not develop cognitive faculties *so that* humans could or would acquire true beliefs. The naturalist story is very different. Our ancestors just so happened to have developed the various cognitive faculties they did because, like every other trait, they improved their reproductive fitness. With respect to cognitive faculties, evolution didn't have true belief in mind (because, not having a mind, it has nothing in mind). So naturalism did not (because it could not) evolutionarily shape our cognitive faculties so that we might more amply get in touch with the truth. Naturalism is neutral with respect to whether our cognitive faculties would or could produce true beliefs. Our cognitive faculties, on the naturalist assumption, did not evolve so that they would produce true beliefs. Traits evolved that better suited us for reproductive success than those possessed by our less fortunate ancestors. That is how the evolution of our cognitive faculties looks from the philosophical perspective of metaphysical naturalism. It is a story that undermines any confidence we might have in our cognitive faculties.

How, on the other hand, might we understand the evolution of our cognitive faculties if God created through evolution? Let us consider the argument of Alvin Plantinga. Plantinga affirms evolution, making it clear that his argument should not be taken as an argument against evolution. Evolution, he says, is not the problem. Naturalism, he argues, is the problem, and with respect to knowledge, theism is superior to naturalism. Plantinga's famous evolutionary argument against naturalism provides the evolutionist who is also a metaphysical naturalist with reason to doubt two things:

(1) that a purpose of our cognitive systems is to serve us with true beliefs; and
(2) that our cognitive faculties do, in fact, provide us with mostly true beliefs.

If the naturalist denies any purpose to the cosmos, then the denial of (1) follows automatically. If there are no cosmic purposes, then our cognitive systems couldn't have had a purpose, let alone the purpose of providing us with true beliefs. So, assuming naturalism, it is not a purpose of our cognitive systems to serve us with true beliefs. Enough said.

But what about (2)? Does the evolutionary naturalist have any reason to believe that our cognitive faculties produce mostly true beliefs? Do they do so, as a matter of fact?

The problem is that unguided evolution (that is, evolution embedded within the worldview of naturalism) cares very deeply for traits and behaviors that are conducive to survival but is utterly unconcerned about true beliefs; again, naturalism is unconcerned because it has no concerns (and no purposes). As naturalist philosopher Richard Rorty argues, evolution does not construct us to "get things right." Natural selection operates on traits or behaviors that enable the organism to move properly, to get its body parts in the right places—for example, getting our mouths open to ingest food while also keeping our body parts out of the mouths of predators. Evolution guarantees only that we *behave* in certain survival-promoting ways; evolution is not much concerned with beliefs. Successful traits, then, are those that help us eat, fight, run away, and have sex. If our cognitive faculties result from naturalistic evolution, their function is simply survival. Given naturalism, then, we have good reason to doubt that their function would be to produce true beliefs.

While behavior may be adaptive, nothing follows about beliefs. Maybe our beliefs are true, maybe they aren't. Philosopher Stephen Stich claims, "Belief formation systems that are maximally accurate (yielding beliefs that most closely approximate external reality) aren't necessarily those that maximize the likelihood of survival: natural selection does not care about truth; it cares only about reproductive success."[6] Our beliefs *might* be true, but evolution plus naturalism gives us no reason to think they would be. So, we *think* our beliefs are true, they appear for all the world to be true to us, but there is no evolutionary reason they would be true. The fault, however, is on the side of naturalism, not evolution.

Evolution and Truth

In his book *The Gay Science*, Friedrich Nietzsche presents a rather grim estimation of human knowledge. He discusses the prospects of gaining the truth with our evolutionarily produced cognitive faculties. He writes: "Over immense periods of time the intellect produced nothing but errors. A few of these proved to be useful and helped preserve the species: those who hit upon or inherited these had better luck in their struggle for themselves and their progeny."[7] Evolution's trial and error is a profound winnowing process, for sure, but—and here is the key point—it retains by definition only what is useful for survival (not what conduces to truth). With respect to our cognitive faculties and the beliefs produced by them, according to Nietzsche, human beings have settled into certain comfortable and characteristic errors that have proved useful for their survival.

These errors, Nietzsche argues, are so deeply ingrained that we now hold them as unquestionable, obvious, or self-evident truth. We take as certain beliefs that are the end result of this happenstance process (our evolutionary inheritance). But if evolution shaped human cognition through a contingent process that aims only at survival (and not at all at truth), then the products of our intellects are, according to Nietzsche, deeply in error. Those deep and systematic errors that form our evolutionary cognitive inheritance are then unquestioningly assumed as paradigm instances of knowledge. Nietzsche writes: "Such erroneous articles of faith, which were continually inherited, until they became almost part of the basic endowment of the species, include the following: that there are enduring things; that there are equal things; that there are things, substances, bodies; that a thing is what it appears to be; that our will is free; that what is good for me is also good in itself."[8] The cognitive faculties that we discussed in the opening chapters—for example, belief in an external world and things that persist through time—aren't, according to Nietzsche, innocent until proven guilty. Our evolutionary inheritance shows these faculties, "such erroneous articles of faith," to be guilty. Although our cognitive faculties may be adaptively advantageous, evolution's sensitivity to reproductive success or adaptive advantage confers no confidence in the truth of their outputs. Evolution's keen sensitivity to survival is, perforce, its insensitivity to truth.

Conceding the consequences of what he takes to be a blind (i.e., naturalistic) process, Nietzsche is left with two options: believe nothing with confidence, or develop a definition of truth that can be altered according to one's own purposes (he boldly affirms the latter). Coming to grips with our evolutionary inheritance, Nietzsche argues, undermines any assurance we might have that our cognitive faculties approach the truth (as traditionally understood). We can have confidence in getting our body parts into or out of the right places, but evolution washes out any grounds for confidence that we have grasped the world aright.

The reason for this has to do with the way the theory of evolution, when combined with naturalism, casts doubt on the reliability of our cognitive faculties. Naturalism, recall, denies the existence of anything other than spacetime and material objects in spacetime. Thus, for the naturalist, there is no Purposer or Designer that exists outside the natural world. As a result, the process of evolution is *unguided*: naturalism does not "care" about the outcome of natural selection and, likewise, truth.

More recently, Patricia Churchland has argued that biology challenges the commonsense notion that the brain's primary function is to acquire propositional knowledge about the world. According to Churchland, from an evolutionary point of view, the "principal chore of nervous systems is to get the body parts where they should be in order that the organisms may survive. . . . Improvements in sensorimotor control confer an evolutionary advantage: a fancier style of representing is advantageous *so long as it is geared to the organism's way of life and enhances the organism's chances of survival.* Truth, whatever that is, definitely takes the hindmost."[9] To translate: evolution, for Churchland, exerts its selective pressures on our ways of thinking and representing to promote survival, with no guarantee that our ways of thinking will produce the truth. Evolution, which is concerned with reproduction and not truth, should cause us to doubt the veracity of our beliefs. She fleshes out Darwin's worry and Nietzsche's skepticism a bit: evolution aims at survival, not at true belief. As Martie G. Haselton and Daniel Nettle put it: "The human mind shows good design, although it is design for fitness maximization, not truth preservation."[10]

Biologist Michael Ghiselin concurs: "We are anything but a mechanism set up to perceive the truth for its own sake. Rather, we have evolved a nervous system that acts in the interest of our gonads, and one attuned to the means of reproductive competition. If fools are more prolific than wise men, then to that degree folly will be favored by selection. And if ignorance aids in obtaining a mate, then men and women will tend to be ignorant."[11] Evolution, to be sure, gives us "the means of reproductive competition," but in so doing it also gives us reason to doubt the veracity of our beliefs. If evolution is true, Ghiselin avers, we should all be skeptics—as noted previously, disbelievers in, say, our beliefs about the past, the external world, the future, other people, and even gods.

But we aren't skeptics. We assume that our cognitive faculties are reliable and, consequently, that the majority of our beliefs are true. We know, or at least claim to know, what time it is now, what we had for breakfast, and whom we will meet for dinner; we assume, in all of this, the external/material world, the past, that the future will be like the past, and that we can come to know other persons. We think that the world—this solid, substantial material world—has existed long into the past and will continue to exist long into the future (and in very predictable ways). We believe that other people exist, and we make plans to befriend them, mate with them, and avoid them. In all of this, once again, we *assume* that our cognitive faculties are working properly—that is, producing mainly true beliefs. We rely on those beliefs to make our way in the world.

But should we trust our cognitive faculties? Are we warranted in maintaining this assumption? Or should we hold that "the human mind shows good design, although it is design for fitness maximization, not truth preservation,"[12] and concede the unsavory consequences of our evolutionary heritage? In short, is it possible that evolution undermines confidence in our cognitive faculties and, hence, undermines our aspirations to knowledge?

Beliefs and Survival

It is hard to imagine that our own beliefs aren't for the most part true; it is difficult to imagine having good reason to think some of our cognitive faculties might be unreliable. But our primate cousins are highly successful at survival without a rich set of beliefs or even without any beliefs at all. Even if primates have some beliefs (and it is increasingly difficult to believe that some, say apes, don't have at least some), then just move a link or two down the Great Chain of Being to any other highly successful species that acts without belief. Except for humans and a few other mammal species, nearly all successful species have evolved behaviors that aid survival without the direction of or appeal to beliefs. Having true beliefs is hardly essential for a species' survival; evolutionarily speaking, it is at best the exception, not the rule.

Even in the human species, beliefs aren't required for most of the traits necessary for our survival. We don't and need not decide to breathe, pump blood, digest food, or grow skin. Nature thought it better to relegate such actions to unconscious and involuntary impulses. And imagine if nature had not. Imagine if you had to decide each time to breathe in just the right amount of air, to squeeze blood from heart to limbs but not too much or too little, and to send acids and enzymes to digest the food that you had just instructed your mouth to chew and your esophagus to transport from mouth to stomach. You could never sleep, and your brain would explode from cognitive overload.

Recent discoveries suggest that it is, in some cases, possible to disassociate human belief and action. For example, some people who have suffered brain damage are nonetheless capable of what is called "blindsight." While blindsighted individuals are unable to see or form certain beliefs about the objects around them, they are able to locate and grasp the objects. They need not be aware of or have beliefs about those objects in order to act. In the 1970s, the neurologist Lawrence Weiskrantz encountered a brain-damaged patient, known as "DB," with some visual blind spots. When shown a pattern of striped lines, DB protested that he could not see them at all. But when Weiskrantz asked DB to guess how many striped lines were vertical, DB answered correctly almost 90 percent of the time. Apparently, his brain was perceiving the stripes while his mind was not conscious of them. [13]

Even healthy brains can lead us to false beliefs. One example of this concerns our beliefs regarding free will. The way most of us think about the process that leads to human action goes something like the following:

"Think first, decide second, act third." However, this common understanding has been called into question by a series of experiments performed in the 1970s by neurophysiologist Benjamin Libet. Libet had participants perform a voluntary act—a simple flick of the wrist—at any time they chose to do so. What he found was that the subject's self-awareness of their wish to act happened about 200 milliseconds prior to the movement of the muscle, which makes perfect sense. However, Libet also found that *even before the subject's self-awareness* of their wish to act, the subconscious brain fired in readiness for the act. This firing happened about 550 milliseconds prior to the movement of the muscle.[14] What these experiments seemed to show is that *unconscious* electrical processes in the brain come first, before *conscious* decisions to perform (what seems to us to be) volitional acts.[15] The feeling of being consciously motivated to act comes *after* the unconsciously moved action; we then think, retrospectively but wrongly, that our beliefs motivated the action—we project our motivation onto the first cause. In Libet-style experiments, beliefs are decorative tagalongs, inessential for action.

I don't know if Libet's sweeping conclusions are correct (indeed, I think they are not), but the point is, they could be. Evolution *could have* fitted us to act first and then to form beliefs (even for those beliefs to be unconnected to the act).

If beliefs are essential for human survival, then we are the only (or one of the very few) species with this trait. But even in the human case, beliefs are inessential for a great deal of our survival traits and behaviors, and, given blindsight and Libet examples, possibly superfluous.

The thought then occurs: Could or would *unguided evolution* (naturalism plus evolution) have produced something so extravagant as true beliefs? Traits and behaviors, no problem—but true beliefs? True beliefs (or many true beliefs) aren't essential for human survival. If they are, we are the only species (or perhaps one of a few) with this trait. Traits and behaviors are clearly linked to survival, but true beliefs aren't. Moreover, the cognitive faculties required to arrive at truth take time and energy. If these beliefs weren't necessary or even particularly beneficial to human survival, why would evolution favor them?

Finally, another reason to believe that unguided evolution (naturalism plus evolution) is unconcerned about the production of true beliefs is the fact that evolution seems to have produced some beliefs or tendencies to believe that are patently false. We have already mentioned two in warmup. Whether due to cognitive malfunction or imprecision, we have developed some characteristic but false beliefs. We have properly functioning, normal tendencies to false beliefs. Like the citizens of Lake Wobegon, we all have a tendency to believe that we are better than average. No doubt overestimating our abilities and intelligence was a good thing for our ancestors in the grasslands of Africa: a self-deceptive boost could keep the optimist going when the pessimist is ready to give up.

Psychologists Hugo Mercier and Dan Sperber have argued that the evolutionary adaptation served by our reasoning capacities is not the ability to find truth.[16] Instead, these capacities evolved to allow us to win arguments and convince others of the views we hold. The evolutionary function of reasoning, they contend, is persuasion, not truth. While the arguments for this claim are somewhat speculative, if true, this would add even further evidence to the position that unguided evolution is unconcerned with truth.

Bottom line: evolution, when it finds its home in naturalism, pulls the rug from under our hopes that our cognitive faculties do any more than help us survive.

The Expectation Principle Applied

One might think that the case so far—that unguided evolution would not lead us to expect true beliefs—is based on examples of people being predisposed to having false beliefs. It is not my contention that unguided evolution would predispose us to having false beliefs. I am only trying to show that it is possible for us to live, act, and survive perfectly well with false beliefs or even no beliefs at all. I gave these examples to loosen up our nearly insurmountable intuition that we need true beliefs to survive. I believe that we have mostly true beliefs, but, like fish and flatworms, we could have evolved to survive without the guidance of true beliefs. Even if we have beliefs, we could have evolved so that (a) they aren't true and (b) they don't function to help us survive. But most of us, me included, think that we evolved true beliefs that help us survive. So these examples help us think about *what if*.

These examples, of course, don't establish that unguided evolution would not lead to our beliefs being generally true. We *may* have acquired true beliefs through a process of unguided evolution. We have not refuted that possibility. But even if this is *possible*, is it *likely*?

Naturalism, when combined with evolution, leads us to expect nothing about the acquisition of true beliefs. It leads us to expect that species that superseded our primate ancestors would have different, even more effective, survival behaviors and traits, but beliefs may or may not be relevant to survival. Naturalism does not care a whit for true beliefs, primarily because naturalism cares for nothing at all. Unguided evolution leads us rationally to expect nothing about our cognitive faculties and the truth or falsity of their outputs.

However, if there is a God, then we might expect that God would create us with cognitive faculties that produce significantly true beliefs. If we are created in God's image and God is a knower, then we might expect humans to be knowers as well. And if our greatest good is found in community with each other and with God, then we might expect beliefs that conduce to our greatest good: moral and spiritual beliefs that help us live together in community with one another and with God.

So, to use the expectation principle once again, let us take as data:

D = Evolved human cognitive faculties are successful at gaining the truth.

And let us take supernaturalism and naturalism as the competing hypotheses:

S = Supernaturalism: there is a being with sufficient powers who is interested in creatures who are capable of gaining the truth.

N = Naturalism: nothing exists but spacetime, and material objects and events in spacetime; there are no supernatural beings.

Since one would expect D much more given S than given N, one has much more reason to believe S than N. We have good reason, therefore, to prefer supernaturalism over naturalism.

If we are, by and large, capable of grasping the truth, which worldview would best explain this? Our cognitive faculties find their intellectual home in the worldview of supernaturalism, while the worldview of naturalism is intellectually inhospitable to reliable cognitive faculties.

The theist can and should recognize the characteristic ways in which we often get things wrong (indeed, most theists have historically emphasized human fallibility; we are, after all, creatures and not gods). But the theist, unlike the naturalist, has an independent reason to be confident that, despite these examples, we are still able to arrive at truth and should not be skeptical about the majority of our beliefs.

We aren't skeptical about the reliability of our cognitive faculties, nor are we skeptical about the truth of many, perhaps most, of our beliefs about the world. It does seem that, overall, we are capable of grasping truth. I know when I am holding a book in my hands, I recognize my daughter when she walks into the room, and I believe that Paraguay really exists when I see it on a map. Which worldview best explains the reliability of our cognitive faculties? Theism gives us reason to believe that humans have cognitive faculties that are reliable and do arrive at important truths. Naturalism, on the other hand, leads us to expect nothing about our cognitive faculties. Further, when we combine naturalism with evolution, we find good reasons to doubt the reliability of our cognitive faculties. Thus, the worldview of naturalism seems intellectually inhospitable to accepting the reliability of our cognitive faculties.

Complications

Suppose we assume that our cognitive faculties were formed by unguided evolution. Under that assumption, we need to look at the proportion of our beliefs that are true compared to those that are false. If most of our ordinary beliefs are true, it follows that unguided evolution in general leads to true beliefs. I have argued that theism but not naturalism provides us with independent reason to think we would have roughly true beliefs. But does not naturalism also have such an independent reason—namely, that, in general, for those creatures capable of belief, false beliefs will hurt survival? Showing that in some instances this is not true does not undermine this more general claim. A complete defense of the evolutionary argument against naturalism would need to rebut this claim.

Further, evidence that we have propensities to form false beliefs in some areas cuts against theism: if God created our minds, then God created with belief-forming mechanisms that systematically produce false beliefs. Again, a more thorough defense of this argument must consider this objection.

This problem multiplies. Consider examples that might tell against the theist's premise that God cares if we have true beliefs on important matters. For example, if God is so concerned that we arrive at true beliefs about important matters, why are there so many widely divergent religious and moral beliefs throughout the world? They cannot all be right.

The theist, to fully defend this argument, needs to develop a reply to these counterarguments. A full and complete defense of the evolutionary argument against naturalism would have to offer an explanation of unreliable belief-forming mechanisms, including, and perhaps especially, those that produce false religious and moral beliefs. I shall simply concede these problems and move ahead without addressing them.

The Initial Likelihood of Theism

Let me not overstate what I have argued. I began the chapter with the thought that it is good to find evidence in support of one's beliefs even if they are immediately justified beliefs. Then I argued that our reliable cognitive faculties provide just that sort of supporting evidence for theistic belief. I have not refuted naturalism, and I have not rejected evolution. I have only argued, using the expectation principle, that, given the truth of evolution, one's reliable cognitive faculties strongly confirm one's already-held belief in God over its major intellectual competitor, naturalism.

Suppose you had just heard a similar argument "confirming" a friend's belief in the existence of ghosts. Suppose you are eating a late dinner at your friend's house, and you hear a loud, inexplicable howl coming from one of the bedrooms. Your friend tells you not to worry—the noise is just from the ghost of a person who was killed in that room at precisely the day of the week and time that you heard the howl. In fact, you learn, you could hear that howl every week at exactly the same time. Thinking the existence of ghosts very unlikely, you scoff. "No, really," your friend insists. "It is a ghost. We completely sealed off the room with caulk, so we know it is not the wind. We had a plumber fix the pipes so we know it is not the plumbing. We had an exterminator drive out all of the animals, so we know it is not rodents."

By the time your friend is done, he has eliminated all of the hypotheses that you had considered as more likely or plausible explanations. The only one left, so it seems, is the ghost hypothesis, and, if it were true, you would expect to hear that strange noise. Your friend, then, demands that you agree with him—the evidence, he contends, is incontrovertible. "You must believe!" he shouts.

Are you, as your friend insists, required by reason (by virtue of the expectation principle) to accept the ghost hypothesis? I think not—even if the evidence is startling and even if you cannot think of a plausible alternative explanation. The problem is that the expectation principle is not the only rational principle that is relevant to your judgment. You are also obliged to attend to judgments of antecedent likelihood. And you don't believe in ghosts. The ghost hypothesis, even if you lack a better explanation, is not on the table.

Although the expectation principle would lead you to expect the data, your judgment about the likelihood of ghosts independent of the evidence weighs very heavily against the ghost hypothesis. In fact, since most readers of this book are strongly disposed to disbelieve in ghosts, our judgment about the initial unlikelihood of ghosts completely and rationally settles the matter. The evidence, even in the absence of a viable alternative, simply is not strong enough to persuade our unbelief.

Is the argument that I offered confirming belief in God similar to your friend's argument for ghosts? Does my argument suffer from similar problems of initial implausibility? Should the God hypothesis, like the ghost hypothesis, be nipped in the bud at the outset? I think not. Or, better, it depends on the person.

Consider the difference. Your friend has attempted to persuade you that there is a ghost in the house. Regardless of your assessment of initial plausibility, your friend thinks the evidence requires you to believe in ghosts. I have not offered an argument that attempts to persuade all readers that they should believe in God. I have only offered an argument for the person who already believes in God, one who has not ruled out God at the outset, one for whom God is a living option. For that person, I have argued that our truth-conducive cognitive faculties can confirm or strengthen their initial degree of belief. I have not argued that the unbeliever is now under some obligation to believe. I have argued only that the thoughtful religious believer—the one who is thinking about knowledge, God, evolution, cognitive faculties, and evolutionary psychology—can find, in the neighborhood, evidential assurance that her belief is true.

The non-theist, the naturalist, on the other hand, thinks that the initial likelihood of theism is really low, maybe even impossible. Given her initial unbelief, nothing I have said should be expected to make theism rationally viable (to her).

But the theist needn't be troubled by the naturalist's judgments here. While the naturalist's judgments about the initial likelihood of God's existence may rationally settle the matter for her, they don't settle the matter for those with different assessments of the initial likelihood of God's existence. For those who are inclined toward God's existence, the evolutionary argument against naturalism may rationally push them from agnosticism to theism or, more relevant to our discussion, may confirm their already-held theistic belief.

Back to the beginning: belief begins with trust. Most of us, anyway, don't believe in other persons or the external world as the simplest hypotheses, or as scientific hypotheses, or as explanations based on experiential evidence (at least not of the technical sort valued by philosophers). We don't believe in other persons or the external world as a scientific hypothesis or explanation at all. We simply find ourselves with those beliefs (due to in-built cognitive mechanisms, ones we now hopefully understand better than when we began this book). Yet every smile and every tear, every mountain and every sea confirm these deeply held convictions. For those who have found themselves believing in God, their trust can both survive and even embrace the evolutionary psychology explanations of religion. If God did create humanity, then religious belief is not reducible without remainder to psychic urges. And if persons of faith believe they know things, their faith can be more assured (given the truth of evolution) than it was before they read the argument in this chapter.

Evidence independent of the evolutionary argument against natural-ism for God's existence may be found in the cosmological argument or the ontological argument. What's more, consciousness, morality, and the ap-preciation of beauty can be used to further support the existence of God. And though the history of believers who have claimed to experience the divine cannot prove God's existence, their testimonies serve as evidence that strengthen the prior likelihood of the hypothesis being true.

Finally, if there is a God, then every joy and all of life confirm the be-liever's already deeply held conviction. This is a poetic way of saying that the world of the theist is a very different world than the world of the non-theist. Lest I wax too eloquent, I must concede that suffering can and does provide disconfirmatory evidence. If I were writing a longer book or a book that was not focused on evolutionary objections to belief in God, I would need to say more about disconfirmatory experiences or evidence.

Belief in God

The post-Darwin intellectual landscape has changed, and it has changed us forever. Our theological heritage located humanity just below the angels. That may still be true. But we must now concede that we are also just above the apes. We are human animals; but we are, to be sure, animals. As Darwin soberly reminds us:

> We must, however, acknowledge, as it seems to me, that man with all his noble qualities, with sympathy which feels for the most debased, with

benevolence which extends not only to other men but to the humblest living creature, with his god-like intellect which has penetrated into the movements and constitution of the solar system—with all these exalted powers—Man still bears in his bodily frame the indelible stamp of his lowly origin.[17]

The human mind likewise bears the indelible stamp of its lowly origin. Our "godlike intellect" has "penetrated into the movements and constitution" of the mind and the forces that have shaped it. We may be much more than the sum total of our evolutionary pushes and pulls, but we aren't less than the sum total of those pushes and pulls.

Cognitive and evolutionary psychology suggest that the same pushes and pulls that drive us to mates and away from enemies likewise attract us to gods and ghosts. Belief in gods is the very natural and very ordinary expression of our very natural cognitive faculties. And yet it is this very ordinariness, this very naturalness that shakes our convictions and makes us fear that God is no more tenable than ghosts. We want to believe that we are special and that there is something unique and special about our religious beliefs. The God-faculty, like the opposable thumb, is at one and the same time nothing special and, if theism is true, truly remarkable. But we fear the God-faculty might be just one more rustic tool for rutting around on the earth, inadequate for exploring the heavens.

As we bear within us the indelible stamp of our lowly origin, we are slowly, ever so slowly, learning what it means to be creatures. We aspire to be intellectual gods, with intellects towering imperiously over the world, but we are in and of this very world we seek to comprehend and even master. We are, then, not gods, intellectual or otherwise; and, being creatures in and of this world, created from its very dust, we must use, as best we can, our dusty cognitive equipment to comprehend all of reality.

Our dusty cognitive equipment has been shaped over eons by a process of trial and error that aimed, if it aimed at all, at reproductive success—whatever served to get us into those loving arms and keep our limbs out of that hungry mouth. Not how I would have done it. But, then, I was not asked. So while we are good at counting a few sticks that might be fashioned into a shelter, we aren't so good at deducing all of the decimals of pi or grasping infinity. We can easily see lions and tigers and bears, but not inside atoms or the center of the sun. We look to the stars to set the seasons, but we can scarcely imagine distant galaxies and wormholes. And we can grasp just enough of other persons to mate, make friends, build alliances, and avoid enemies; but we have not come up with the recipe for world peace. Our dusty cognitive equipment is pretty good at some rather mundane activities, but the rest is a stretch.

And yet we have calculated pi out to the *n*th digit, we have delved deep into the atomic structure of reality, we have transported our minds to the edge of the universe, and we get along vastly better now than we did just 10,000 years ago. Dusty equipment and all, we have put it to some remarkable and unexpected uses.

So how about taking the agency-detecting device and the theory of mind all the way to the gods? Is that on a par with electrons and dark matter? Or is it more like everyone believing they are above average or in Santa Claus?

I have given reasons to trust the God-faculty in spite of the claims of those who would explain God away. And I have offered evidence—our reliable cognitive faculties—to confirm belief in God. I have not offered an argument that would require religious belief for everyone, or that would prove non-theists irrational, or that would settle the matter of God's existence once and for all. I have argued only that a religious believer could find her belief strengthened by the argument presented in this chapter.

Believers and unbelievers alike do the same thing: we use our evolutionarily shaped cognitive equipment to do the best we can to understand the world, a world that does not easily yield its mysteries. Our intellects are considerably less godlike than our ancestors thought, and considerably more animal-like than most of us willingly concede. When branching out beyond mating and hunting and gathering (and a little elementary arithmetic) and into philosophy and physics, we are liable to error. Agreed. And so we should open ourselves to the comments and criticisms of those who disagree with us. As stated in the first chapter: we all want the truth. It is just that the truth is sometimes really hard to figure out.

CHAPTER 6

Atheism, Inference, and IQ

Ripped from the Headlines

Philosophers could save a lot of time by taking their cues from the news. In recent years headlines have proclaimed that science has demonstrated the rationality of unbelief and the irrationality of belief: "Logic Squashes Religious Belief, A New Study Finds."[1] The venerable *Scientific American* proclaimed: "Losing Your Religion: Analytic Thinking Can Undermine Belief."[2] And then there is the alleged correlation between high IQ and atheism. While God may run in our genes, unbelief, so it is claimed, is the exclusive province of the intelligent. The *Guardian Liberty Voice* trumpeted, "Atheists More Intelligent Than Religious Believers Says New Study,"[3] while *The Independent* preferred the condescending converse: "Religious people are less intelligent than atheists." In the veritable *Medical Daily* we read, "Proved: Atheists More Intelligent Than Religious People."[4]

In this chapter, we will examine some of the cognitive underpinnings of atheism. While the cognitive science of religion is well-trodden ground, atheism has been considerably less scrutinized. I suspect this is due to the following: the vast majority of those who work on these topics are atheists or agnostics who view religious belief as false, outdated, and even bizarre. Given this assumption, the project of socio-psychological critiques of religion is to explain how otherwise rational people could hold obviously false beliefs. Unlike religious belief, their own beliefs (agnosticism or atheism), so the narrative goes, are products of rational reflection—the triumph of reason over superstition. The project, then, is to seek out the malfunction that produces religious beliefs; atheism gets a free pass.

There is no lack of anecdotal evidence to support this narrative. Many cognitive scientists, whose work is, in principle, neutral with respect to the truth of religious belief, betray a firm belief in the rational superiority of atheism. A cursory glance at various practitioners' conceptions of religion reveals their stance: religion is patently improbable and factually impossible,[5] an airy nothing.[6] One might not have thought there were so many ways of saying "false." What could lead the human mind to entertain and accommodate obviously false beliefs? The narrative continues. Given the pathetic absurdity yet superabundance of God-beliefs, non-truth-tracking belief-producing mechanisms must be the culprit. With the discovery of these ignoble cognitive causes, religious belief can be finally unmasked as the irrational folly we all know it is. Atheism, on the other hand, is not the product of ignoble causes; it is the result of cool reflection on reasons. Religious belief is debunked because it is produced by non-truth-conducive psychological causes, whereas unbelief is rational because it is based on rational reflection on good reasons.

In this chapter we will consider claims suggesting that atheism is a virtuous intellectual achievement (atheists are smarter, atheists are more inferential, etc.), whereas theism is intellectually suspect.[7] And then we will consider various cognitive defects that are correlated with atheism. We will ask at various points along the way, What follows, about atheism and theism alike, from these fascinating studies?

Atheism and IQ

In the essay "Average Intelligence Predicts Atheism Rates across 137 Nations," Richard Lynn's team's analysis of many individual studies shows that intelligent people tend to be less likely to believe in God.[8] The data come in many shapes and sizes. For example, in Western societies, high IQ correlates with low belief in God, and scientific elites are considerably less likely to believe. As children grow up and grow in intelligence, they are less likely to believe. During the twentieth century, as IQ increased, religious belief declined. Finally, nations with a higher average IQ have higher numbers of atheists. Lynn says that one factor alone explains atheism: "I believe it is simply a matter of the IQ." While there is some reason to doubt the correlation between higher intelligence and a tendency toward unbelief,[9] even supposing it true, the question remains: What is the most plausible explanation of this correlation? Does high IQ turn people into atheists?[10]

Some atheists have taken this correlation to confirm the triumph of reason over superstition. The atheist is comforted: my peers are smarter, more analytic, more rational, more evidence-sensitive, more truth-concerned than theists. Lynn's team affirmingly quotes Sir James Frazer's assertion in

The Golden Bough that, as civilizations developed, "the keener minds came to reject the religious theory of nature as inadequate . . . religion, regarded as an explanation of nature, is replaced by science." And they begin their essay with the following: "Dawkins' recent book *The God Delusion* suggests that it is not intelligent to believe in the existence of God. In this paper we examine the evidence for this contention." Has Lynn established that "the keener mind" is the driver of unbelief?

Suppose there is a correlation between atheism and high IQ. Would the correlation between intelligence and unbelief show that intelligent reflection on the evidence was the cause of unbelief? There are other possible and even plausible explanations of the data.

The most plausible theory suggests that the common cause of both higher IQ and unbelief is increased socioeconomic status. We know this much to be true: IQ levels increase as material wealth increases, and (with the US as a notable exception) belief in God has decreased in Western nations as material wealth has increased.[11] Moreover, within nations, those who are higher up socioeconomically likewise tend toward unbelief. If there are existential urges to religious belief—ones exacerbated by starvation, sickness, and death—then societies that can satisfy those urges through economic advances are likely to see a decline in religious belief (and, as health and access to education improve, a corresponding incline in IQ). Rational reflection, however, is notably absent from this explanation of unbelief. Lynn's paper concedes this possibility but then ignores it. But as Phil Zuckerman writes: "One leading theory comes from Norris and Inglehart, who argue that in societies characterized by plentiful food distribution, excellent public healthcare, and widely accessible housing, religiosity wanes. Conversely, in societies where food and shelter are scarce and life is generally less secure, religious belief is strong."[12] The satisfaction of social needs, not critical reflection on the God-hypothesis, is the best-attested explanation of religious unbelief in modern society. If Zuckerman is right, both atheism and theism are mediated by existential anxieties (and their relief).[13] Since relieving existential anxiety is not truth-conducive, the atheist is not in a rationally privileged position over the theist.

What about the decided lack of belief in God among academics, in general, and highly accomplished scientists, in particular? According to Lynn, only 7 percent of members of the American National Academy of Sciences believe in God and only 3.3 percent of the fellows of the Royal Society believe in God. Lynn claims that such high-IQ people, as they grow up and reflect on -religious beliefs, slide into doubt and unbelief. Isn't this evidence that high

IQ (which Lynn takes as shorthand for "people with highly educated critical skills who deeply reflect on all of their beliefs, not just their scientific ones") produces reflective unbelief?

Once again, correlation does not prove cause. There may be underlying biases that nonreflectively incline academics to unbelief. Such biases are often unconsciously activated; later on, though, when considering the beliefs produced by biases, we offer up a rational justification for that belief (and sincerely assume that we acquired that belief on the basis of those good reasons). Although academics fancy themselves immune to the normal psychological biases that affect everyone else, they aren't immune to such biases and consequent rationalizations.[14] Let me suggest one bias that likely affects the prevalence of atheism in the academy: conformity bias.

If you find yourself in Rome doing as the Romans do (say, sipping a late-night espresso after eating a delicious gelato—which, given both your diet and your aversion to caffeine before bedtime, you'd never do at home), you have probably succumbed to conformity bias. Humans have a powerful, innate, and subtle tendency to conform to social norms. Probably no surprises here: we subconsciously conform to, for example, various beauty standards, including the latest fashions in clothing or hairstyle. We don't typically choose in a self-conscious way to conform just as we don't typically decide for ourselves what's beautiful or cool. Our culture tells us what is beautiful and cool, and we unselfconsciously accede. When we are at the store, we simply find ourselves coveting that new style of sweater or the latest cell phone, even to the point of buying them.

While conformity usually affects behaviors, we also conform with respect to beliefs. Through a series of studies, Solomon Asch showed how difficult it is for an individual to maintain her own belief in the midst of a group that expresses a contrary opinion.[15] In one experiment, subjects were shown a picture of a target line along with an array of lines of three different sizes, identified as (a), (b), or (c), and were asked which of the three lines was the same length as the target line. This simple perceptual task had an obviously correct and easily verifiable answer—say, (c)—which the subjects would choose correctly. However, when all of the other members of the group insisted that the correct answer was, say, (a), 75 percent of the subjects changed their belief to the wrong answer. Conformity increases when the issue involved is less obvious and when other members of the group are of a higher social status: when individuals view the others in the group as more powerful, influential, or knowledgeable than themselves, they are more likely to conform to the beliefs of the group.

We find precisely this situation in the academy. If the majority in a highly desirable group holds certain views or manifests certain practices, then you should expect aspiring candidates for that group to mimic those beliefs and practices. As countercultural as professors sometimes can be, you don't see many male professors with mullets or wearing kilts. And if the majority of the scientific community is atheistic or agnostic, then you should expect to find belief conformity (better, unbelief conformity). Those who aspire to be members of that community will find themselves mimicking the unbelief of the majority on the basis of acceding to this conformity disposition (and not on rational reflection). The cognitive science of religion does not claim that everyone is inclined to a faith that they would die for. It only claims that, given certain cultural influences or other environmental influences, people are inclined to easily acquire and sustain religious beliefs and practices. Religious belief may be widespread and skin deep. And so people might be inclined, due to other cognitive dispositions and in the face of other cultural or environmental influences, to unbelief. Believers and unbelievers alike may have acquired their religious belief/unbelief through a nonreflective, non-truth-conducive cognitive process, not on the basis of rational reflection.

One might wonder how unbelief gets started in the first place, and one might think that, because we are talking about scientists or professors, it must be through rational reflection. But more mundane explanations are on offer. As the Christian pop tune rightly notes, it only takes a spark to get a fire going. What are those first sparks of atheism? All it takes is a famous atheist scientist here and there (perhaps with religious belief driven underground due to the rise of methodological naturalism[16] or because of a mean Sunday school teacher) to get the fire started. Conformity bias typically takes hold in the wake of prestige bias. Prestige bias, manifesting in an unconscious preference for high-status individuals, is likewise likely a factor in the inculcation of unbelief. In highly competitive academic communities, people knowingly do all sorts of things to ingratiate themselves with and gain access to a "famous" scholar. If famous scholars self-identify as unbelievers, the influence on junior and aspiring scholars mounts. Again, acceding to prestige and conformity biases is not a conscious process. One likely does not even know that that is why one believes as one does.

The social pressures against religious belief in academia aren't insubstantial. A prominent social scientist friend was an adult convert to Christianity. He kept his convictions mostly private, and he never referred to them in his scholarly work or during presentations. But when word of his conversion leaked out, he was routinely asked by colleagues at meals or even during public presentations if the stories of his conversion were true (usually followed by a snide comment, asking how he could believe something so ludicrous). His colleagues and the audience would laugh as he did his best to offer a response. Imagine the effect of the public derision of this respected scholar's religious beliefs on students or younger scholars. They got the message.

Expressions of unbelief needn't be overt. From the selection of anti-religious texts for course assignments to the haughty raising of an eyebrow, affirmations of unbelief are abundantly clear. Students learn what questions can be asked and what cannot, what can be said and what should not. A well-timed horse laugh, a whispered comment over beers after a lecture, and a derisive snort are much more effective than an extended argument in a scholarly journal. In short, when prestige bias combines with conformity bias, unbelief in the academy is to be expected.[17] When a majority of people holds a certain belief, conformity bias tells us how difficult it is for nearly everyone to resist the urge to conform. Pretty soon, with respect to contrary opinions, it is horse laughs all around.

Recall the correlation between high IQ and unbelief. Academics have considerably higher IQs on average than the general public, and academics are considerably more likely to be atheists or agnostics. Is this correlation due to smartness on the part of academics? If I am right, a major contributor to unbelief in the academy is not intelligence; rather, unbelief is at least partly the result of the unconscious urge to conform.

Of course academics won't think that their religious unbelief is due to an unconscious psychic urge. (Theist to atheist scholar: "You just reject belief in God because you want so desperately to be accepted by that group.") No one does. And yet we are all—academic and non-academic, high and low IQ—susceptible to unconscious psychological urges. None of us is immune from biases, including the drive to conform.

Let me provide one more example of conformity bias in the academy that lends some credence to my claim about religious belief. A 2010 study showed a correlation between political liberalism and high IQ, on the one hand, and political conservatism and considerably lower IQ, on the other. It found that liberals' average IQ score is 106.42, whereas conservatives' average IQ score is 94.82.[18] Are liberals, then, smarter than conservatives? Is liberalism thereby more likely to be true? A recent survey suggests that only 11.5 percent of university faculty identify as politically conservative, whereas 62 percent identify as politically liberal or far left.[19] If a disproportionate number of university professors are politically liberal, then a disproportionate number of high IQs are going to skew politically liberal. And if a disproportionate number of university professors are politically liberal, then we should likewise expect prestige and conformity bias on the part of their students that would result in their becoming politically liberal. And, since those with university education have, on average, higher IQs than those without, we will find liberalism correlated with, but not necessarily caused by, higher IQ.

What is the best explanation of political liberalism in the academy? Careful and reasoned assessments of the arguments pro and con? I doubt it, and for two reasons: (a) arguments in political philosophy aren't sufficiently compelling to expect rational consensus in favor of political liberalism, and (b) most academics don't spend their time carefully assessing arguments in political philosophy (they just do their biology or anthropology). What might contribute, then, to the predominance of political liberalism in the academy? Most likely, conformity and prestige bias (with some hiring bias to boot). If most professors are liberal, IQ scores will skew toward the liberal (but not for truth-conducive reasons).[20] Moreover, if a disproportionate number of professors are politically liberal, then their students (who will have higher IQs than those who don't attend college) will, due to conformity bias, tend toward political liberalism. The correlation between high IQ and political liberalism, then, is explained (at least in part) by prestige and conformity bias, not by intelligence or careful attention to the arguments.

Some academics may insist that they are less subject to conformity biases than the general public. Perhaps so. After all, it is reflection and careful thought—the sort of things that professors do for a living—that help us discover new ways of thinking and acting. So, while there may be some conformity bias that accounts for part of the difference, maybe that is not the whole story. Let us grant that professors are less subject to conformity than the average person. But therein lies another account for their greater-than-average atheism: if breaking away from ordinary thinking is a mark of the highly intelligent because they can override conformity (at least in some cases), and if ordinary thinking is theistic, then some may reject theism by virtue of being nonconformist. A similar dynamic may be at play in people with really high IQs being more likely to believe that there is no external world, that causation is an illusion, that minds aren't real, that humans aren't free, and that there are no moral truths.

Moreover, this entire discussion accepts that professors have higher IQs than other professionals. Although professors have higher IQs than the average person, we don't know whether they have higher IQs than other professionals (who are much less likely to be non-theists).

Even if there is a correlation between high IQ and unbelief, it does not follow that intelligence or rational assessment of arguments mediated or contributed to anyone's unbelief. Unbelief could be the result of such unconscious belief drivers such as the relief of existential anxieties or conformity bias.

Atheism and Inferential Thinking

If religious belief is culturally recurrent, natural, and intuitive (noninferential), then we should expect unbelief to be relatively rare, non-natural, and inferential.[21] Studies have shown a correlation between inferential thinking and unbelief. Consider the ABC News headline "Religious faithfuls lack logic, study implies."[22] Nicholas Epley claims that religious belief is "more of a feeling than a thought."[23] Faith is a matter of heart, unbelief a matter of mind. In *Scientific American*, we read about "How Critical Thinkers Lose Their Faith in God" (theists, presumably, are uncritical thinkers).[24] Rob Brooks, Scientia Professor of Evolutionary Ecology at UNSW Australia, claims that it is becoming "clearer that religion is, in some senses, the opposite of rational thinking."[25] Do such studies show, as these headlines assert, that atheists are rational but theists aren't? Do they demonstrate the rational superiority of atheism over theism?

Atheism is indeed relatively rare. Only about 2.4 percent of American adults identify as atheist (67 percent of these are men—more on that in the next chapter).[26] A 2012 WIN-Gallup poll puts the figure worldwide at 13 percent.[27] Prior to the twentieth century, the figure would likely have been near zero.

How about the naturalness of atheism? If, as CSR suggests, belief in God is as natural as enjoying music, then atheism seems as unnatural as hating it. But it is hard to know if this or that belief is natural or unnatural without having some sense of what it means for a belief to be natural.

Let us move into an understanding of "natural" by way of analogy: religion seems natural in the same way that language is natural. Humans have the inborn capacity to naturally develop a language if they are in the right sorts of environments. Being raised by wolves in a forest is not the optimal setting for language acquisition; being raised in a human community anywhere in the world is. Give a child time, and her inborn cognitive dispositions will issue forth in fluent Chinese, say, or Turkish. She won't decide to acquire a language; native language acquisition will happen naturally, with little or no conscious effort on her part.

Likewise, we are cognitively disposed as we mature to acquire the religious beliefs of our family or culture with little or no cognitive effort on our part; they just arise, naturally. While we are cognitively disposed to religious beliefs and practices, which religious beliefs and practices one acquires depends entirely on one's culture (substitute equals for equals in this sentence and the same is true for languages). Religion, like language, is natural—it is an easy, straightforward, unconsciously acquired expression of universal cognitive faculties in the right environment.

Unbelief, in this sense, is not so natural, or at least not so clearly natural. Of course, if one grows up in a community that lacks typical religious promptings, one's natural dispositions to religious belief aren't likely to be appropriately stimulated (and thus one is not likely to acquire religious beliefs and practices). The Swedish child of a Swedish atheist (or two) who grows up in Sweden is likely to naturally be a non-theist. Naturally occurring religious beliefs require a religiously infused environment to flourish, just as the natural acquisition of the Swedish language requires the right sort of environment to flourish.

Even if religious belief were natural and unbelief unnatural, there is nothing special about a belief's being natural: a belief's being natural is neither a virtue nor a vice. A natural belief can be rational or irrational, true or false, a brain fart or a revelation. For example, an entomologist who is fully aware that only about a dozen of the tens of thousands of species of spiders are poisonous might still have a natural (but, given her awareness of the data, irrational) fear of spiders. On the other hand, I might naturally and rationally believe that I am better than average (this is something each of us is naturally disposed to believe), and it might even (in some sense or other) be true. However, once I have been made aware that all humans have the cognitive disposition to think themselves above average (and rightly judge that we cannot all be correct in making such judgments), then my belief would be irrational. But if I were unaware of this cognitive disposition, or if I didn't realize that we cannot all be right in making such judgments, then my belief would be, as far as I can tell, rational. I could multiply examples. What makes beliefs rational or irrational, or true or false, has nothing to do with their being natural.

Nonetheless, some beliefs—the natural ones—arise rather easily and without conscious reflection in virtually everyone due to our cognitive constitution, given the right sorts of stimulation.

What about atheism? If humans typically, through no inferential effort of their own, believe in God, some people might need to reason, with some inferential effort, their way to unbelief.

Consider an analogy with folk physics, which, like religious belief, is culturally recurrent, natural, and intuitive.[28] Folk or naïve physics is our unreflective, perceptual understanding of the physical world. Folk physics might include simple and true generalizations such as "Dropped rocks fall to the ground" and "Rocks thrown hard enough at windows will break them." It also includes commonsense statements that run contrary to contemporary physics, which postulates a host of unobservable entities such as atoms and photons (and may even hold that our natural notions of past and future are illusory). The movement from folk physics to contemporary physics required an enormous amount of inferential effort, effort sufficient to override at least some of our deep and natural intuitions.[29] Contemporary physics, requiring abstract thinking and complicated mathematics, is deeply counterintuitive and contrary to what we observe. Belief in contemporary physical theories, then, requires inferential thinking. Likewise, the rejection of our very natural religious beliefs may involve inferential thinking.[30]

Just this sort of reasoning guided Will M. Gervais and Ara Norenzayan through a series of studies to determine the effect of inferential (what they called "analytic") thinking on religious belief and unbelief.[31] Gervais and Norenzayan offered a series of inferential prompts to determine their effect on religious belief and unbelief. They hypothesized that inferential thinking would override one's more natural and intuitive cognitive inclinations toward religious belief. Because the headlines mentioned above relied on their studies, I will consider them in some detail.[32]

In the first study, using the cognitive reflection test developed by Shane Frederick, they offered three problems.[33] Their study will make more sense if you stop and think through your own response to the problems before proceeding to their analysis. The problems are as follows:

1. A bat and a ball cost $1.10 in total. The bat costs $1.00 more than the ball. How much does the ball cost? ___ cents
2. If it takes 5 machines 5 minutes to make 5 widgets, how long would it take 100 machines to make 100 widgets? ___ minutes
3. In a lake, there is a patch of lily pads. Every day, the patch doubles in size. If it takes 48 days for the patch to cover the entire lake, how long would it take for the patch to cover half of the lake? ___ days

In each case, the quick and easy intuitive response is incorrect, while the more deliberate inferential response is correct.[34]

Participants were then measured with respect to religious belief and unbelief, responding to statements such as the following:

> In my life I feel the presence of the Divine.
> It does not matter much what I believe as long as I lead a moral life.
> I believe in God.
> I just don't understand religion.
> God exists.
> The devil exists.
> Angels exist.

Gervais and Norenzayan found that success on the cognitive reflection test was negatively correlated with affirmations of religious belief; inferential thinking, they claimed, was negatively correlated with religious belief. So, in their terms, Gervais and Norenzayan concluded: "This result demonstrated that . . . the tendency to analytically override intuitions in reasoning was associated with religious disbelief."[35]

A second set of studies involved unconscious primes, with a series of prompts designed to elicit inferential thinking. For the sake of illustration, let us consider just one, the disfluency prime.[36] Disfluency primes involved fuzzy fonts rather than the large, clear fonts of the control group. Gervais and Norenzayan's claim is that having to figure out a fuzzy font engages inferential thinking in a way not required when reading large and clear fonts. The subjects again completed a measure of belief in God/religiosity. Again, Gervais and Norenzayan concluded that the set of studies reinforces the hypothesis that inferential processing decreases religious belief.

More recent studies affirm the hypothesis that if religious belief is more intuitive and noninferential, then unbelief should be a product of inferential reasoning. Amitai Shenhav, David G. Rand, and Joshua D. Greene conducted a cognitive reflection test study similar to that of Gervais and Norenzayan with over eight hundred participants (US residents) with a median age of 33; intuitive responses were positively correlated with religious belief and inferential responses with unbelief.[37] Their two other studies combine with this one to show a correlation between intuitive thinking and belief in God and inferential (analytic) thinking and unbelief.

Gordon Pennycock, James Allan Cheyne, Paul Seli, Derek J. Koehler, and Jonathan A. Fugelsang sampled over two hundred people across the

United States with a median age of roughly 35.[38] They measured inferential thinking style (again, which they called "analytic cognitive style" or ACS) in two ways: first with a variant of the cognitive reflection test, and second with base-rate conflict (BRC) problems (problems that contain a conflict between a stereotype and probabilistic information). Since religious engagement is likely correlated with religious belief, they measured belief according to an individual's reported level of participation in, for example, church and prayer. They also measured religious beliefs through one's degree of belief in heaven, hell, miracles, the afterlife, angels, demons, and an immaterial soul. Finally, they queried participants about what kind of God, if any, they believe in: answers ranged from theism to atheism. While the researchers produced many nuanced results, overall they affirmed the intuition that inferential thinkers are more likely to be unbelievers than intuitive thinkers. The first study, for example, offered evidence of "an analytic [inferential] tendency to ignore or override initial intuitive responses."[39] They concluded that inferential processing decreases the likelihood of supernatural belief.

Other Atheisms

Gervais and Norenzayan are aware that not all atheists are inferentialists (and not all theists are anti-inferentialists). They have identified, in addition to analytic (inferential) atheism, at least four additional types of atheists. Perhaps the most interesting is mindblind atheism, seen in individuals who lack the ability to mindread—that is, to process or cognize other minds.[40] Such individuals are usually on the autism spectrum, lacking to various degrees what we have called theory of mind (see the next section). The higher one is on the autism spectrum (that is, the less one is able to cognize persons), the less likely one is to believe in a personal God. If theistic beliefs involve belief in a divine person, those who lack the ability to cognize persons are likewise likely to lack belief in a divine person (more on this at the end of this chapter).

Some atheists, which Gervais and Norenzayan call "inCREDulous," simply lack adequate and relevant cultural inputs to form religious beliefs. Religions typically involve costly and observable credibility-enhancing displays (CREDs), such as fasting, tithing, chastity, and even martyrdom.[41] Such hard-to-fake actions, which signal one's commitment to cooperate with ingroup members, seem to contribute to the credibility and acceptability of a religion within a culture.[42]

In a culture lacking CREDs (for example, in Northern European cultures), one might find religious belief simply incredible. Apatheism typically results from an indifference to God that arises from existential security; apatheists are content with their existential security and as a consequence don't care much about belief in God.

I have already discussed the influence of conformity and prestige bias on atheism. In addition, not mentioned by Gervais and Norenzayan, some atheists deeply want God not to exist. Jean-Paul Sartre, for example, did not want the kind of cosmic authority that an omniscient judge would have. He didn't want there to be a God who knew his every thought and action (and judged him accordingly).

Just as there are psychic, cultural, and non-truth-conducive impulses involved in religious belief, so, too, there are psychic, cultural, and non-truth-conducive impulses involved in unbelief. So one cannot infer from the claim that atheists as a group have a more inferential thinking style that any particular atheist acquired her belief inferentially.

Atheism and Autism

In the previous sections we learned of the correlation between atheism, on the one hand, and, on the other hand, a cognitive virtue (inferential thinking) and a cognitive excellence (high IQ). Gervais and Norenzayan assert that the connection between inferential thinking and atheism is more than a correlation. While conceding alternative explanations, they claim that inferential reasoning "explains all of these findings in a single framework that is well supported by existing theory regarding the cognitive foundations of religious belief and disbelief."[43] In other words, when inferential processes are engaged, they trump/correct/erode/block intuitively and uncritically assumed religious beliefs. This surely happens in some cases: since research suggests that we are cognitively inclined to form beliefs in gods and spirits, atheism may emerge when these natural beliefs are subjected to criticism by or overruled by inferential thinking, rightly or wrongly. But perhaps in some cases the relationship of inferential thinking (and even high IQ) to unbelief is not the relationship of cause to effect; perhaps both are mediated by a common, underlying cause.[44]

I have already suggested some common causes of these correlations—conformity and prestige biases. Let us, in this section, consider one more: autism. In this section, I will explore features of autism that possibly mediate both atheism and inferential thinking (and high IQ). I will present studies that suggest that the connection between atheism and inferential thinking is mediated by mentalizing constraints in autism.

Studies have shown that atheism and agnosticism are, in some cases, both correlated with and mediated by mentalizing constraints, which are common to individuals on the autism spectrum. If God is conceived of as a nonphysical mind, then those with mentalizing constraints should be expected to manifest higher degrees of unbelief in a personal God and, again in some cases, atheism.[45]

The work of Catherine Caldwell-Harris and colleagues first called attention to the connection between high-functioning autism (HFA) and atheism.[46] Their research suggested that individuals with high-functioning autism are predisposed toward unbelief. Norenzayan, Gervais, and Kali H. Trzesniewski argue that the mentalizing deficits characteristic of HFA individuals incline them toward atheism; while they replicated the Caldwell-Harris study, which showed that those higher on autism scales are less likely to believe in a personal God,[47] they were also able to isolate and then eliminate other cognitive faculties or tendencies that might mediate or incline one toward atheism.[48]

The first part of the Caldwell-Harris study involved a content analysis of blog posts on religion and philosophy in websites (a) for HFA individuals (wrongplanet.net) and (b) for non-HFA (neurotypical) individuals (golive wire.com/teen). Posts were coded for content analysis, and individuals were assigned to various religious belief categories, including, for example, agnostic, atheist, and Christian. Participants with HFA were significantly less likely to code for theistic beliefs and significantly more likely to code for atheism and agnosticism. For example, while about 37 percent of the neurotypical population coded for Christianity, less than 17 percent of the HFA population did. HFA discussants were 50 percent more likely to identify as atheists and 70 percent more likely to identify as agnostics.

A second study involved an internet questionnaire given to 61 individuals with a self-identified autism spectrum condition; 105 undergraduate students composed the neurotypical control group. Participant location on an autism scale was determined through three diagnostic tests, and they self-assessed their religious belief.[49] While HFA and neurotypical individuals were equally likely to be agnostics (about 14 percent), HFA individuals were more than three times as likely to be atheists (34 percent vs. 10 percent). Moreover, while over 50 percent of neurotypical individuals were Christian and Jewish, only 28 percent of HFA individuals identified as Christian or

Jewish. Finally, the higher a person scored on the autism spectrum quotient, the more likely she or he was to be an atheist. In brief, the results of the internet questionnaire confirmed the results of the blog post content analysis. But while Caldwell-Harris and her colleagues confirmed a connection between HFA and atheism, they were unable to determine the source of that connection.[50]

Norenzayan's studies sought to isolate mentalizing deficits as the mediator of unbelief. Based on scores on the autism spectrum quotient self-reporting questionnaire, Norenzayan and colleagues claimed that mentalizing deficits mediated increased tendencies toward atheism and agnosticism. If God is personal, then a typically functioning ToM seems associated with belief in God as an intentional agent. Mentalizing deficits therefore seem a likely contributor to disbelief. They concluded: "Cognitive approaches to religion emphasize that a reliably developing social cognitive mechanism—mentalizing or theory of mind—is a key foundation that supports the intuitive understanding of God or gods. Present findings bolster this hypothesis, and further demonstrate that mentalizing deficits undermine not only intuitive understanding of God, but belief as well." If this is the case, autistic individuals with more severe mentalizing difficulties may be cognitively incapable of conceiving a God that neurotypical individuals might intuitively recognize as personal and intentional.[51]

If Caldwell-Harris and Norenzayan are correct, atheism's mediation by autism provides an explanation of the correlation between atheism, on the one hand, and high IQ and inferential thinking, on the other. Since autistic individuals (as a group) have both a higher IQ and more inferential thinking styles than non-autistic individuals (as a group), autism provides the common ground between atheism, on the one hand, and high IQ and inferential thinking, on the other. In a sufficient number of cases to explain the differences, atheism and inferential thinking may be mediated by autism (rather than atheism being typically mediated by inferential thinking). Given the relatively slight differences between believers and unbelievers, a more careful analysis of the data correlating unbelief, on the one hand, and higher IQ and inferential thinking, on the other hand, may be explainable almost entirely by the prevalence of individuals on the autism spectrum, whether diagnosed or diagnosable as autistic or not, in the relevant studies.

Finally, autism is correlated with being both an academic and a scientist.[52] As a group, women are the least likely to be on the autism spectrum, while men are four to five times more likely than women to have autism. Professors, as a group, are more likely to be autistic than men as a group, and scientists are even more likely. This scale intuitively maps onto religious sensibilities. Women, as a group, are considerably more religious than men, men than professors, and scientists than professors.[53] If this mapping is correct, autism would partially explain the atheism/agnosticism we find both in the academy and among scientists.

Conclusion

The claim that atheists are smarter, more attentive to arguments, and hence more rational than theists has not survived scrutiny. Atheists may have higher IQs and more inferential thinking styles than theists, but that does not make them more rational. Indeed, high IQ, inferential thinking, and atheism, one and all, may share non-truth-conducive common causes. Since high IQs cluster around universities, the prestige and conformity biases that shape universities explain some of the preponderance of atheism in universities. And increased wealth both leads to high IQs and relieves the existential anxieties that undergird some religious unbeliefs. Finally, autism is highly correlated with both atheism and agnosticism, on the one hand, and high IQ and inferential thinking, on the other. Both are mediated by the autistic individual's mentalizing deficits. In short, while some people surely reject God's existence after careful consideration of arguments, unbelief can be accounted for in many cases with the same sort of psychological biases that some attribute to theists. It is complicated, of course. But this much we know: atheism and agnosticism are associated with the same sorts of psychological biases that atheists attribute to theists.

CHAPTER 7

Atheism, Autism, and Intellectual Humility

You Just Believe That Because, Revisited

We have reached a point in the book where we have probably learned just enough cognitive science to be dangerous. Atheist and theist alike are sufficiently lightly armed to start slinging mud at those who disagree with them; we should also have noticed that slingers of mud are likely to get dirty.

We can imagine, then, the atheist sophomore philosophy major (or Richard Dawkins) declaring that theists are irrational, stupid(er), and deluded, and that atheists, on the other hand, are rational, smart(er), and more evidence-sensitive.

Alternately, we can imagine the theist sophomore philosophy major declaring that lurking beneath the atheist's higher IQ and inferential thinking are all sorts of sordid cognitive biases and defects—prestige and conformity biases, for example, and, the clincher, autism.

Both, theist and atheist alike, may now think the other's belief a delusion, the product of an inbuilt irrationality mechanism.

We might imagine the following conversation:

Atheist: "You just believe in God because of an inbuilt irrationality mechanism."

Theist: "By irrationality mechanism, do you mean agency-detecting device and theory of mind?"

Atheist: "Yes."

Theist: "But those are both perfectly ordinary, natural cognitive faculties. Belief in God is ordinary and natural."

Atheist: "Ordinary and natural, agreed. And wrong."

Theist: "I concede I might be wrong about God's existence. It's hard to know for sure if some of our beliefs are true."

Atheist: "It's not so hard to tell. In fact, the smartest people in the world are atheists. They can tell you."

Theist: "What do you mean the smartest people in the world are atheists?"

Atheist: "Well, studies show that those with the highest IQs are atheists."

Theist: "That may be, but it doesn't mean that you're smarter than I am, does it?"

Atheist thinks to himself, "Yes," but says: "I guess not."

Theist: "Isn't higher IQ associated with wealth, and doesn't wealth relieve some of the existential anxieties that undergird at least some people's religious belief?"

Atheist: "I suppose. I hadn't thought of that."

Theist: "Maybe you just reject belief in God because your existential anxieties have been relieved, not because you're smarter than I am. Of course, it's *natural* to believe on the basis of the relief of some psychic urges, but that doesn't make your belief rational. It's natural to believe that you're a really good person, thus relieving your existential anxieties, but that doesn't make your moral self-assessment rational. Of course, I'm not saying that you aren't a good person."

Atheist: "I'm sure my unbelief isn't due to something so trivial. Atheists not only have higher IQs; we also have more inferential thinking styles. We're used to basing our beliefs on careful assessment of evidence and arguments. Theists, however, not only have lower IQ; they have more intuitive, impulsive thinking styles. I'm an evidence assessor, and that's why I think I'm rational and you aren't. You just believe in God because you aren't as smart or as sensitive to the evidence as we atheists are."

Theist: "So *you* are smarter and more reflective than *I* am?"

Atheist thinks to himself, "Yes," but says: "Well, er, I mean, on average atheists have higher IQs and more inferential thinking styles. Not sure where you and I fall."

Theist: "Do you really think that if I were smarter and relied more on inference, I would see things the way you do?"

Atheist: "Yes, of course. The smartest people in the world, university professors and scientists, are also the least religious."

Theist: "So, you think professors and scientists are atheists because they have some special insight and are especially attentive to the arguments for and against the existence of God?"

Atheist: "Yes, professors and scientists are, after all, well trained in logic and science. So they're more inclined to base all of their beliefs on evidence."

Theist: "Yes, but in their training, didn't they pick up on all sorts of clues and cues from their teachers that belief in God is bunk? And, since they're eager to find their place in the profession, don't you think some of their religious beliefs simply melt away, unnoticed, without any rational reflection whatsoever? Professors and scientists just believe God doesn't exist because they've succumbed to social pressures. Professors and scientists are people, too, you know. They aren't immune to the biases of less-well-educated people. They may be smarter and they may be more inferential, but the cause of their unbelief is conformity bias pure and simple."

Atheist: "Well, I concede it is possible that some professors and scientists are atheists because they have complied with social standards, but most of them just look coolly and rationally at the universe and infer that God does not exist."

Theist: "Cool and rational reflection, huh? You may be right about cool. Don't studies show a connection between atheism and autism? And aren't professors and scientists considerably more likely to have some sort of autism?"

Atheist: "I hadn't heard that, but it does make some sense. My profs weren't always the best at recognizing and responding to social cues."

Theist: "That is one sign of high-functioning autism. The underlying cause seems to be some sort of mentalizing deficiency. Autistic individuals aren't very good at judging other people's thoughts, feelings, or desires. And so they don't know how to respond appropriately. Here is my point about religious belief—if there is a personal God who is communicating through his creation, a person with a mentalizing deficiency is not likely to get it. Professors and scientists just believe there is no God because they cannot, because they are autistic."

Atheist: "Maybe that is true of some, but they do have higher IQs and better thinking skills."

Theist: "Right. But those could also be caused by their underlying autism. Autistic individuals have higher IQs and are more inclined to think inferentially. Autism is a common cause of all three—higher IQs, inferential thinking, *and* atheism. Professors and scientists aren't atheists because they are smarter; they are atheists because they have a cognitive defect."

You get the "You just believe that because . . ." point. We have a deplorable tendency to explain another person's beliefs and behavior according to an underlying psychological or epistemological defect. Of those who disagree with us on important matters, we often find ourselves saying (or at least thinking to ourselves), "They're crazy!" Of course, we never attribute our own beliefs and behaviors to such defects—we consider ourselves paradigms of rationality.

Read any political editorial—on welfare reform, say, or human-made climate change—and see how easy it is for the author to think that her views are obviously well-supported by the evidence *and* that those who disagree with her are ignoring key pieces of evidence, are unable to follow simple logical reasoning, are in the pocket of corporations, don't care about poor people, and are just plain deluded. She can barely conceal her scorn. And if she can conceal *her* scorn, read the anonymous comments below the editorial, comments rife with unconcealed scorn, derision, and invective. The comments on both sides reveal our attitudes toward those who disagree with us: they aren't only wrong; they are irrational, immoral, and crazy.

But I, if I am normal, demur—*I* am not like *them*. I am the Cartesian free spirit who dispassionately floats above the socio-historical fray, deducing theorems from self-evident axioms. I am an epistemic god. The riff-raff beneath me are mere socially conditioned, embodied creatures subject to subterranean cognitive passions and malfunctions.

Sadly, though, I, like them, am not very good at knowing what really informs my beliefs and motivates my actions. More importantly and even deeper, I, like them, am not the independent, free, self-conscious, rational reflector that I believe myself to be. Aristotle may have called humans rational animals, but we are more animal than rational.

While we may think we have acquired our belief/disbelief in God through our individual awareness of, deep reflection on, and considered endorsement of good reasons, none of us is the coolly rational solitary reflector we claim to be. We seldom believe and act only after rigorous and thorough and neutral reflection on the evidence sifted and weighted appropriately. We may reject Descartes's mind-body dualism, but we treat ourselves as though we are minds floating free of the epistemically debilitating influences of bodies, communities, cultures, and histories. We consider ourselves rational calculators who start with bias-free evidence and deduce the truth. I believe my cognitive equipment to be better than that of others (at least well above average) and my grasp of the relevant data clearer. And so, based on my own hard work, I am uniquely situated with respect to the truth. And, if you disagree with me, you are not. You are biased, your critical skills less honed; you are a child of your culture, subject to prejudices or maybe even delusions. I am rational. You are irrational, maybe crazy.

Studies Show

We have muddied the waters considerably. Some of us (heck, all of us), some of the time and in some circumstances, are puppets on strings pulled by implicit biases and desires. Studies show us what some of those are. With respect to religious belief and unbelief, we are influenced by ToM, ADD, conformity bias, and autism, among many others. Does the science of the mind, then, support judgments of irrationality and/or lunacy? What do the studies really show?

All we have learned from these studies is something about groups of people. We don't know anything at all about any individual people. Given what we know about groups, we know that if an individual is autistic, that individual is statistically more likely to be an atheist or agnostic than a neurotypical individual. If an individual is neurotypical, that individual is statistically more likely to be a religious believer than is an autistic individual. Moreover, we also know that an atheist or agnostic is statistically more likely to be autistic than neurotypical and a religious believer is statistically more likely to be neurotypical than autistic. Beyond that, we know nothing at all about any particular neurotypical or autistic individual; nor do we know anything at all about any particular atheist or religious believer. Armed with this information, one simply has no idea if Richard Dawkins is high and Mother Teresa is low on the autism spectrum.

One cannot know in the case of any specific atheist or agnostic if her unbelief is mediated by mentalizing deficiencies (even if we were to know that she is high on the autism spectrum). Indeed, one cannot know the cognitive and cultural influences on any particular individual's belief. From the problem of evil to having had a bad Sunday school teacher or a bad relationship with one's father, there is a host of possible sources of unbelief.

We can illustrate this point with a parallel example. Suppose it is true, not uncontroversially, that men are better than women at math.[1] This claim is typically based on sets of standardized test scores such as the SAT. As recently as 2013, despite efforts to reduce the gap, there was a 32-point male advantage on the math section of the SAT. The differences are more startling at the high end of the scores: over the past twenty years, the ratio of males to females who score in the top 5 percent in high school math has remained constant at two to one.[2] Supposing, again controversially, that these tests mirror a gender difference in mathematical abilities, what does the difference really show? It shows how a group of people performed on an achievement test; it tells us nothing about any individual person whatsoever. It does not show that every man is better than every woman in math. It does not give me license to think, when I see a woman, that I am better than she at math. And if I see a startlingly high math score, I have no reason to attribute that score to a man (or, if I see a low math score, to attribute that score to a woman). How groups of people perform on a test tells me nothing about the performance of any individual member of that group. Finally, since most scores mostly overlap, men and women are, by and large, roughly equal in math abilities.

Judging any particular person based on this generalization ignores the contingent particularities that shape every individual test taker—their native cognitive abilities (and disabilities), their schooling, their socioeconomic status, their parents (and their parents' education), gender biases in classrooms,[3] how hard they worked in math classes, their personality (temperament). Any individual performance on an achievement test reflects a unique and unknowable combination of native ability, culture, and personality. We are in no position to sort out, in any particular case, whether or not cognitive ability alone determined an individual's test score. And we are in no position to infer back to native cognitive ability (alone) based on a test score.

Back to religious belief/unbelief. What we have learned about groups tells us nothing about the psychological mechanisms involved in any particular individual's unbelief. Since we know nothing about the causes of any particular person's unbelief, we know nothing about that person's rationality or irrationality. Lacking access to any particular individual's consciousness, we must be agnostic about the sources of her unbelief and, therefore, her rationality or irrationality.

Debunking Narratives

The atheistic debunking narrative holds that religious beliefs are irrational because they are caused by unreliable cognitive mechanisms, whereas atheism is rational because it is the product of rational reflection on true beliefs.

The alleged connection between atheism, on the one hand, and inferential thinking and high IQ, on the other, offered narrative support. We have both complicated and debunked a portion of the narrative: atheism is correlated with and (sometimes) mediated by a mentalizing deficit; it is not (always) the product of rational reflection. Theism, on the other hand, is produced or shaped by properly functioning and reliable cognitive faculties, including at least ToM and ADD. With respect to psychological causes, belief in God seems superior to atheism or agnosticism. But we should resist both theistic and atheistic debunking narratives.

We simply cannot know, in any particular case involving any particular individual, the cognitive processes involved. Even supposing that mentalizing deficits are implicated in some atheistic beliefs and the God-faculty is implicated in some theistic beliefs, we still have no idea whether any particular person's belief was produced by these processes. Learning, then, that various cognitive mechanisms are involved in producing and shaping religious belief and unbelief in general tells us nothing whatsoever about the beliefs of any particular believer or unbeliever.

The brain is vast and mysterious. We have only begun to plumb its depths. Human culture is likewise vast and mysterious, and reducing cognitive functioning to fixed brain processes *inside* people means neglecting what takes place outside, *between* people. Outer impacts change our inner worlds. Theistic belief and unbelief are thus likely the result of both complex cognitive processes and cultural influences. Determining, in any particular case, the precise influences on anyone's religious belief or unbelief is, at least for now, impossible.

With respect to the rationality of atheism and agnosticism, Norenzayan, Gervais, and Trzesniewski offer wise counsel: "We emphasize that our data don't suggest that disbelief solely arises through mentalizing deficits; multiple psychological and socio-cultural pathways likely lead to a complex and overdetermined phenomenon such as disbelief in God."[4] This is surely the conclusion to be drawn. Such dynamics are also necessary to include in autism studies, embracing the fact that people on the autism spectrum, too, are individuals.

Let us now complete our debunking of the atheistic debunking narrative. Substitute "belief in God" for "disbelief" and "the God-faculty" for "mentalizing deficits" and we get: "We emphasize that our data don't suggest that belief in God solely arises through the God-faculty; multiple psychological and socio-cultural pathways likely lead to a complex and overdetermined phenomenon such as belief in God." This is surely the conclusion to be drawn about religious belief as well. We cannot see inside another's mind to see either their beliefs or their mental shortcomings. Humility, not arrogant pronouncements on another's character or beliefs, seems the order of the day.[5]

My approach from now on will be decidedly different from that of the debunking skeptics who think they have lighted on the psychic urges that induce religious belief. If parity were to prevail, one might argue that atheism is mediated, not by inferential reasoning, but by an underlying cognitive defect common to autistic individuals. Such simple inferences are neither psychologically, sociologically, or epistemologically plausible or illuminating. Instead of reducing another person's religious beliefs to primitive psychic urges (to which one does not believe oneself liable), we will simply try to understand two autistic individuals' religious beliefs and practices. We will seek understanding, then, not declarations of rationality or irrationality.

Understanding Autism

In the best-selling novel *The Curious Incident of the Dog in the Night-Time*, the narrator, Christopher, introduces himself to the reader in the opening chapter as follows:

> My name is Christopher John Francis Boone. I know all the countries of the world and their capital cities and every prime number up to 7,057.
>
> Eight years ago, when I first met Siobhan, she showed me this picture [of a sad face] and I knew that it meant "sad," which is what I felt when I found the dead dog.
>
> Then she showed me this picture [of a happy face] and I knew that it meant "happy," like when I am reading about the Apollo space missions, or when I am still awake at 3 a.m. or 4 a.m. in the morning and I can walk up and down the street and pretend that I am the only person in the whole world.
>
> Then she drew some other pictures . . . but I was unable to say what these meant.
>
> I got Siobhan to draw lots of these faces and then write down next to them exactly what they meant. I kept the piece of paper in my pocket and took it out when I didn't understand what someone was saying. But it was very difficult to decide which of the diagrams was most like the face they were making because people's faces move very quickly.

When I told Siobhan that I was doing this, she got out a pencil and another piece of paper and said it probably made people feel very [wavy-mouth face] and then she laughed. So I tore the original piece of paper up and threw it away. And Siobhan apologized. And now if I don't know what someone is saying, I ask them what they mean or I walk away.[6]

Throughout the novel, Christopher is consistently unable to judge what another person is thinking or feeling based on either vocal intonation (sarcasm, for example) or facial expression. Christopher's remarkable inferential and mathematical skills combined with his difficulties in grasping the thoughts and feelings of the other characters suggest that he has a form of autism (which is never explicitly stated in the novel). Christopher is simultaneously socially challenged and mathematically gifted.

Diagnoses of autism often focus on difficulties in understanding social and nonverbal communication (such as body language and facial expressions) and difficulties in understanding linguistic ambiguities (such as metaphors or intonation).[7] Simon Baron-Cohen's (1995) term "mindblindness" is often used to describe the inability to "mindread" what other people believe, feel, or desire.[8] Conversely, neurotypical people are intuitive mindreaders (attributing beliefs, feelings, and desires to human agents rather successfully).

Autism, as one might imagine, comes in degrees. One test of where one falls on the autism spectrum, the "reading the mind in the eyes" test, is similar to Siobhan's "test" of Christopher. This simple test assesses how well one can "read" the emotions of another person simply by looking at photos of their eyes. You will better understand the test if you stop reading and quickly take the test.[9] For any given set of eyes, one is asked to select, from among four choices, the emotions expressed: desire, say, or anger, or surprise, or embarrassment.

There are some fairly typical and universal facial expressions, and humans, as a group, are pretty good at "reading" the emotion or thought behind those expressions. Mind readings aren't inferred—you just look and see, as it were, the thought or emotion in the facial expression. However, individuals with autism have a difficult time accurately assessing another's thoughts and emotions. Such mentalizing difficulties are very likely related to deficiencies in abilities related to the theory of mind (ToM).[10]

Autism presents with a range of symptoms from a complete inability to communicate with others to moderate difficulty in judging other people's feelings. A person with mild autism might be relatively successful on the "reading the mind in the eyes" test, whereas a person with severe autism might be completely unsuccessful. So autism is a spectrum, from mild to severe, with some autistic individuals able to function very well in society and others not so well.

High-functioning autistic individuals also typically have an urge to systemize, where "systemizing is the drive to analyse and explore a system, to extract underlying rules that govern the behaviour of a system; and the drive to construct systems."[11] Systemizing is aided, cognitively, by an excessive attention to detail and a narrowed focus of interest. Nick Dubin and Janet Graetz propose that individuals on the autism spectrum may have a capacity for *logical creativity*, or *a stereoscopical view on the world*,[12] sometimes in ways that fuse scientific and religious reasoning. Extraordinary systemizing abilities are found among those with unique talents in developing highly abstract systems such as physics and mathematics. Systemizing may also manifest in, for example, taxonomies, geography, and astronomy, which can bring understanding and order to an obsession with lizards, mountains, and planets. Some autistic children are obsessed with machines, from burglar alarms to vacuum cleaners, and their inner workings; little wonder they go into engineering.[13]

While systemizing is a superb cognitive tool for understanding physical systems, it is a decidedly ineffective tool for understanding and negotiating the very unpredictable behavior of individual people. For that one needs a fully functioning theory of mind.

Finally, high-functioning individuals on the autism spectrum are associated with average or above-average intelligence; for example, such children score higher on mathematics and physics tests.[14]

The ability to mentalize[15] facilitates an understanding of another person's beliefs and desires (and so facilitates responding to other persons appropriately and empathetically). In the cognitive science of religion, as we have noted, this ability is identified as the basic tool for attributing agency and intentionality, not only to humans, but also to gods, spirits, and other invisible entities.[16]

If properly functioning mentalizing is necessary for theistic belief, its malfunction may preclude believing in a personal God. Could God, conceived of as a person, be inconceivable to an individual with autism? And what effect might the drive to systemize have on one's religious beliefs and practices? Let us proceed in the next two sections by way of example.

Autism and Religious Belief: Temple Grandin

Temple Grandin, subject of an award-winning film,[17] is both autistic and an autism advocate. At the age of two, Grandin was diagnosed with autism, believed to be the result of brain damage. She overcame enormous odds and social stigma to become professor of animal science at Colorado State University, an animal rights activist, and a best-selling author. Her social discomfort led her to find solace among animals, and she has devoted her life to the discovery of painless methods of killing farm animals.

She writes about her religious beliefs in her autobiography, *Thinking in Pictures* (so titled because she thinks in full-color pictures: "When somebody speaks to me, his words are instantly translated into pictures"). Her spiritual journey lies in between her opening self-description, "As a totally logical and scientific person . . . ," and her concluding hope for "one pure moment of silence." Between her propensity to systemize and her hypersensitivity to noise, we find a profound self-description of an autistic spirituality.[18]

What sort of spirituality does one find in a person whose "thinking is governed by logic instead of emotion"? We might ask more pointedly, what sort of spirituality might one find in a high-functioning autistic individual, one with a high IQ, a deficient theory of mind, an urge to systemize, and a sensitivity to noise?

Grandin was raised in a Christian home with daily prayers and weekly church and Sunday school. Already by age eleven or twelve it seemed illogical to her that any particular religion was better than any other at communicating with God and inculcating and motivating moral values. While religion, for Grandin, is "an intellectual rather than an emotional activity," she feels most religious, a sense of awe even, at church services with beautiful organ music and harmonic chanting. Music, in turn, emotionally connects her to childhood memories. She writes:

> Music and rhythm may help open some doors to emotion. Recently I played a tape of Gregorian chants, and the combination of the rhythm and the rising and lowering pitch was soothing and hypnotic. I could get lost in it. There have been no formal studies on the effect of music, but therapists have known for years that some autistic children can learn to sing before they can talk. Ralph Mauer, at the University of Florida, has observed that some autistic savants speak in the rhythm of poetic blank verse. I have strong musical associations, and old songs trigger place-specific memories.[19]

In high school, Grandin began to conceive of God, with no less sense of awe, as an ultimate ordering force for good, one pitted against the destructive forces of the second law of thermodynamics, rather than as an anthropomorphic person. While, like her hero Einstein, Grandin does not believe in a personal God, she considers herself religious and religion to be a very important part of life. Her religious outlook is characterized by a deep sense of trust, hope, and gratitude. Science, she thinks, does not and cannot provide all of the answers to life's questions, including meaning, morality, and mortality. Religion, on the other hand, can and should motivate our highest human goodness. Until 1978, she also had a firm belief in the afterlife.

In 1978, Grandin swam through a noxious vat of chemicals (organophosphates) as a publicity stunt. The toxic effects of the chemicals on her brain were startling. She writes:

> The feeling of awe that I had when I thought about my beliefs just disappeared. Organophosphates are known to alter levels of the neurotransmitter acetylcholine in the brain, and the chemicals also caused me to have vivid and wild dreams. But why they affected my feeling of religious awe is still a mystery to me. It was like taking all the magic away and finding out that the real Wizard of Oz is just a little old man pushing buttons behind a curtain.[20]

The chemicals initially washed away all religious feelings. Later, her sense of awe and occasional feelings of being close to God returned, but her belief in the afterlife did not.

After her chemical bath, reason/science led her back to her religious beliefs. Quantum physics holds that a gazillion subatomic particles, entangled and vibrating and affecting other particles, both near and far, are constantly interconnecting and interrelating, creating new and higher forms of reality. She writes, "One could speculate that entanglement of these particles could cause a kind of consciousness for the universe. This is my current concept of God." A more "scientific," less personal deity is not uncommon among autistic individuals. Jesse Bering writes: "In the autobiographies of autistic individuals, God, the cornerstone of most people's religious experience, is presented more as a sort of principle than as a psychological entity. For autistics, God seems to be a faceless force in the universe that is directly responsible for the organization of cosmic structure—arranging matter in an orderly fashion, or 'treating' entropy—or He has been reduced to a cold, rational scientific logic altogether."[21] Autistic believers are more likely to reject both organized religion and traditional theological formulations in favor of their own rationalized system of beliefs.

Grandin's deep moral convictions concerning the suffering of animals combine with her belief in an emergent, quantumly entangled, moral-ordering force (God) to establish in her a sense of a cosmic consciousness: "Doing something bad, like mistreating an animal, could have dire consequences. An entangled subatomic particle could get me. I would never even know it, but the steering linkage in my car could break if it contained the mate to a particle I disturbed by doing something bad. To many people this belief may be irrational, but to my logical mind it supplies an idea of order and justice to the world."[22] This emergent moral-ordering force exercises a kind of cosmic providence, rewarding the good and punishing the bad.

One finds in Grandin's spiritual autobiography just one way in which autism's cognitive style finds religious expression. Mentalizing constraints inform an impersonal God, systemizing prefers a God as an ordering force for good, and, given Grandin's aversion to loud noises, she seeks solace in silence.

Finally, though just hinted at here, her deficient ToM renders it difficult for Grandin to endure the human contact and distractions one might find in worship services. Grandin finds her sacred space, ironically, among animals being led to their death, animals she feels called by God to love in their final moments. Grandin is noted for designing more humane ways of, not to put too fine a point on it, slaughtering livestock. Grandin's picture-thinking allows her to see this process as an animal sees it, and her hypersensitivity to stimulation allows her to experience the sights and sounds along the way as the animal might experience them (cattle, for example, are sensitive to the shifting of light). Taking the "cow's eye view" and then alleviating their suffering is motivated by her empathic bonds with animals, bonds she lacks with humans.[23] Her technological innovations are designed for the animals to go to their death calmly and peacefully.

It is at this mortal moment when she feels most aware of God. She writes: "When the animal remained completely calm I felt an overwhelming feeling of peacefulness, as if God had touched me. I did not feel bad about what I was doing." At this intersection of life and death, Grandin feels for animals what she cannot feel for humans: a deep and powerful empathy that moves her to kindness and compassion. She feels their peace as her peace. Through leading animals to a peaceful death, her religious feelings returned. "For the first time in my life logic had been completely overwhelmed with feelings I did not know I had."[24]

We may understand Grandin better if we situate her sense of sacredness within a broader autistic framework. We know that she loves animals and that she, like cows, is hypersensitive to light and sound. We know that she hyperfocuses on certain details while ignoring all of the rest. And we know that, as a systemizer, she needs to meaningfully order those often very messy details. She creates a sacred space in, of all places, a slaughterhouse, then. At every level of planning, she is guided by her remarkable ability to see as the cow sees. By cutting out the extremes of darkness and light, she soothes both the animals and herself. In her redesigned squeeze chutes, cattle can breathe normally and not bruise or injure themselves. Her serpentine ramps prevent cattle from being spooked by workers or the slaughterhouse ahead. Out of the chaos of feedlots and slaughterhouses and death, she has created a stress-free, quiet, and even peaceful order. The quantum particles that constitute Grandin become entangled with the quantum particles that constitute her cows. Out of the chaos emerge goodness and peace. There, in that pure moment of mind-ordered silence, is her sacred space.

Autism and Religious Belief: Daniel Tammet

On Pi Day 2004 (March 14), Daniel Tammet, a high-functioning autistic savant, recited pi to 22,514 decimal places. Tammet, "the Brain Man," claims that he read through all of those digits just once and then recited pi from memory. You, on the other hand, might do your level best to recall from your school days that pi is 3.1415 and then be done with it, memory exhausted. How was Tammet able to memorize pi out to more than 22,000 decimal places at a single glance? Tammet sees numbers as colors and shapes.[25] He says, "The number 1, for example, is a brilliant and bright white, like someone shining a flashlight into my eyes. Five is a clap of thunder or the sound of waves crashing against rocks. Thirty-seven is lumpy like porridge, while 89 reminds me of falling snow."[26] When asked to do a major calculation, Tammet can beat an electronic calculator. How does he do it? "When multiplying, I see the two numbers as distinct shapes. The image changes and a third shape emerges—the correct answer. The process takes a matter of seconds and happens spontaneously."[27] When learning pi, he transformed that massive set of digits into a beautiful and memorable (presumably large) landscape, and then reproduced his digital tapestry on demand.

Daniel is likewise astonishingly capable of learning languages. He learned to speak Icelandic, one of the world's most difficult languages, in just one week, culminating in a remarkable demonstration of his fluency on national television. During his week of study, his mind seemed to vacuum up words, absorbing grammar and syntax along with them.

Daniel is also a Christian with religious beliefs and practices reflective of high-functioning autism. While he concedes difficulties in empathic thinking, the sort of thinking associated with belief in a personal god, this has not prevented him from intellectually exploring life's ultimate questions. Tammet concedes that, as with many autistic individuals, his "religious activity is primarily intellectual rather than social or emotional."[28] While he grew up with little interest in religion, he was persuaded of the truth of Christianity through the writings of G. K. Chesterton, an English polymath and defender of Christianity in the early part of the twentieth century. It is hard to say just what in Chesterton moved Tammet, but Chesterton somehow explained the Trinity—that God is a deeply loving and living relationship (Father, Son, and Holy Spirit)—in a way that made sense of reality for Tammet. Like pi to 22,000 decimal places, he could picture the Trinity in his head, and it all made sense.

I discuss Tammet because of his insights into how individuals with autism might appropriate religion. As one might expect with mentalizing constraints, autistic individuals often find communal religious services difficult to negotiate or even tolerate. Given their social difficulties and language problems, autistic individuals find social reciprocity difficult (e.g., estimating when to enter or finish a conversation) and eye contact disturbing, and many have difficulties reading—and using—facial expressions. Since all of these features complicate social situations, sustained mental effort is required to engage in social interaction. The social dimensions of religion that might appeal to a non-autistic person can be stressful and even demeaning for some autistic individuals. So Tammet is not much of a churchgoer: "I don't often attend church, because I can become uncomfortable with having lots of people sitting and standing around me."[29]

Like Grandin, he likewise finds the music soothing and religiously evocative: "On the few occasions when I have been inside a church I have found the experience very interesting and affecting. The architecture is often complex and beautiful and I really like having lots of space above my head as I look up at the high ceilings. As in childhood, I enjoy listening to hymns being sung. Music definitely helps me to experience feelings that can be described as religious, such as of unity and transcendence. My favorite song is 'Ave Maria.' When I hear it, I feel completely wrapped up inside the flow of the music."[30]

It is, however, important to stress that social difficulties don't necessarily mean that people with autism wish to be alone; many of these individuals express a wish for friendship and inclusion,[31] but being confused by social interaction causes them to withdraw.[32] The constant need to process unpredictable and widespread social cues without a fully functioning ToM can be confusing, exhausting, and stressful. As Tammet writes, "Predictability was important to me, a way of feeling in control in a given situation, a way of keeping feelings of anxiety at bay, at least temporarily."[33]

Since predictability offers a sense of security and control, some autistic individuals appreciate the stability of highly regularized rituals and liturgies. Tammet writes: "In fact, many people with autism do find benefits in religious belief or spirituality. Religion's emphasis on ritual, for example, is helpful for individuals with autistic spectrum disorders, who benefit greatly from stability and consistency."[34] Such services, scripted beforehand and regularly repeated, require a minimum of spontaneous interactions that involve mentalizing abilities. Some find the routine and structure of liturgy socially affirming and empowering. Since individuals on the autism spectrum value sameness and repetition due to sensory overload, it has been suggested that religious rituals could be experienced as calming and comforting.[35] Moreover, socializing with people who share interests reduces anxiety and feelings of inadequacy.[36]

Some autistic individuals view social interactions as mere information exchanges (without interference from facial expression or vocal intonation). They may look off to the side of the speaker, avoiding eye contact at all costs. And yet they aren't ignoring the speaker; they are processing the verbal information as best and as deeply as they can (often quite literally). Given the opportunity, the autistic individual may let loose a flood of information until emptied. As Tammet writes, "I did not understand that the purpose of conversation was anything other than to talk about the things that most interested you. I would talk, in very great detail, until I had emptied myself of everything that I wanted to say and felt that I might burst if I was interrupted in midflow."[37] When he was considering Christianity, he attended a weekly fellowship meeting in which he inundated its members with questions (all the while ignoring their testimonies). He simply wanted information that would answer his intellectual questions about the truth of Christianity, information he found only in Chesterton.

A final aspect, which is related to the social dimension, deals with reduced conformity needs. Engaging in mentalizing means accessing and assessing other people's wishes and needs. Healthy and stress-free communal living often requires conforming to other people's wishes and needs. Religious or spiritual contexts up the ante—there are unseen agents (other minds) to please. In a study assessing the so-called audience effect, Keise Izuma and colleagues note: "Our actions are strongly influenced by our belief that they may be seen and evaluated by others."[38] In a dictator game, participants were asked to donate money, once in the presence of an unknown observer and once on their own. In the neurotypical control group there was an increase in donations when being observed, while there was a *decrease* in donations made in the high-functioning autism group. Autistic individuals may lack an awareness of social reputation. Thus, they may be indifferent to the social reputation benefits attained by socializing and engaging in religious or spiritual groups (which, according to theories on audience effect, are motivational for most people). Perhaps, because of mentalizing constraints, individuals that are unaffected by or unaware of the imaginary presence of invisible agents likewise find it difficult to conform to the wishes of an invisible god.

Such increased individualism may explain why some autistic individuals don't participate in communal worship services or hold widely accepted theological beliefs. Individuals on the autism spectrum may embrace religious beliefs according to their own logic (as did Temple Grandin), rather than conforming to a system of principles prescribed by religious or spiritual groups. Others may instead hold more rigidly to religious doctrines to acquire a sense of coherence.[39] Increased individualism may also help us understand the prevalence of atheism among autistic individuals. Individuals with autism may find religion to be less relevant to their lives because they don't mind losing out on the social benefits that accompany participation in a religious community.

Autism and Atheism

We have considered the religious beliefs of exactly two autistic individuals, Temple Grandin and Daniel Tammet. While Grandin's and Tammet's autobiographies are both fascinating and revealing, two persons aren't representative of the entire group of autistic individuals. Writers of books are generally articulate storytellers with higher than average IQs (present company excepted); they are seldom representative of any group to which they belong. Repeating Grandin's and Tammet's stories entails the risk of contributing to the autistic stereotype of the socially insensitive but brilliant savant and scientist. Ten percent of all autistic people are thought to be savants, so 90 percent aren't; there are many teenagers with autism who find mathematics impossible to understand and who are disinterested in science—considerably more than the other way around. Each autistic individual has her own unique set of cognitive capacities, was raised in her own family within her own culture, with her own set of character- and belief-shaping experiences. There is no typical autistic individual, so there is no typical autistic form of religious belief (or unbelief). If you want to understand any particular autistic individual, you need to ask her for her story.

The stark differences in the religious beliefs of these two individuals who share a common pathology offers us a glimpse into the enormous diversity of both causal factors and reasons for belief—or not-belief—that people could have concerning their religious beliefs. They also begin to illustrate the wide range of religious beliefs that people hold. I could have relayed countless stories, just of autistic individuals and their religious beliefs (or, in a small percentage of cases, their unbelief). I could have told stories of physicists who believe and mathematicians who don't, of female atheists and male theists, of Jewish philosophers who are devout and others who are hostile to Yahweh. And so on and so on. We should proceed cautiously when assuming we know what others believe and why, and how they have arrived at their beliefs.

We have focused on the stories of particular people in this chapter to humanize and individualize both autism and religious belief. Most autistic individuals are religious believers with religious beliefs and practices formed and shaped in ways that are influenced by their cognitive styles. Some autistic individuals are unbelievers; some of that unbelief was likely mediated by a mentalizing deficit. But every person is an individual with her own very contingent cognitive makeup within a very particular social context. So each autistic individual should be treated as an individual, not psychopathologized by being lumped into a group.

I suggest that, since we don't have access to the inner springs and mechanisms involved in *any* individual's cognition, we shift discussions of the rationality of a unique individual's religious beliefs or unbelief away from epistemic and psychic evaluation to sympathetic understanding. And if we wish to understand these individuals, we must ask for and listen carefully to their stories.

CHAPTER 8

Googling God

Google

People have been comparing the mind to a computer for quite a while. According to this analogy, mind-brains and computers are alike in being electricity-powered information processors. Computers process that information, as we know, with both hardware and software. Minds, likewise, are hardwired to process information in various ways (we have universal, innate cognitive faculties that produce similar beliefs across time and among cultures). But minds can also be programmed through experience and culture to process in new and unique ways. The input, processed with either a metallic or meat computer, produces an informational output. But suppose we change the metaphor a bit and compare the mind to something in the neighborhood of a search engine—Google or, better, a Google search.

Let us do a Google search on the topic of this book: "belief in God." When I do that I get first, of course, the Wikipedia entry but then, in order, websites that offer "six straightforward reasons to believe that God is really there," a case for God as revealed in Jesus Christ, a listing of *Huffington Post* blog posts including "When We See the Power of God at Work" and "Prayer and Miracles" and "For Those Who Don't Believe . . . ," and a *Christianity Today* article entitled "Why Believe in God? Ignite Your Faith." And that is just page 1. But who really needs (or reads) more than a few of Google's tendered websites to settle once and for all a debate concerning one of life's perennial issues?

Sarcasm aside, Google is a great help. A simple search of the internet would give us way too much information as well as way too few tools for adequately sorting through that information. There are millions upon millions of pieces of electronic data coded for belief in God stored in various computers and databases all over the world. Suppose your search had returned, in no particular order, every page that included the word "belief" and "God." You'd get, among countless other unhelpful sites, one in which someone exclaimed, "Oh my God, that is beyond belief!" A decent and serious search for the best sites on belief in God, then, would eliminate every idiomatic reference to God—no "goddammits," then, and no "God's gift to women," no perfunctory "and God bless America" at the end of every politician's speech, and no empty references to hurricanes or earthquakes as "acts of God."

Google culls references to "belief" and "God" from among the billions upon trillions of pieces of data (a google, one might say) that jointly make up the web. But, as we have noticed, mere culling would still leave us with a monumentally gargantuan and simultaneously tedious pile of mostly insignificant and disordered data. But, thank Google, they sort out those websites that are worthy of our attention.

Google searches first through every piece of available data, picking out those data that include the union of belief and God (non-idiomatically speaking). That's just for starters; that first step—the one that preserves every meaningful, non-idiomatic use of "belief" and "God"—would still deposit on our doorstep a massive pile of undifferentiated data. There are simply too many references to "belief" and "God" for an unfiltered assessment and presentation of the data to yield any meaningful results.

Google, then, does us a favor—it sorts through this massive pile of mostly useless data, selects what is most relevant, and then lists the "most important" sites first (lots of "mosts"). We lack the time, the skills, and the brain capacity to sort through all of the WWW-archived data relevant to a rational assessment of belief in God. It is simply impossible for any single human being to collect all of those bits of data from, after all, the WORLD-WIDE web, select the ones relevant to the issue of belief in God, and then determine which are worthy of consideration. Google's algorithms do all of that for us. Google starts from that undifferentiated mass of data and seizes upon the data that it deems most useful to the searcher. In the blink of an eye, it culls and filters a gazillion bits of data down to the most relevant few, presenting them in order of importance.

Just a few strokes on your keyboard, the click of a mouse, and Google does it all for you: in milliseconds the data best suited to your inquiry appear on your screen. Now all you need to do is to read the best data ranked in order of importance and make up your mind. What could be simpler?

Google's ease, however, comes with a cost. Google's filtered and ranked information is anything but neutral with respect to the truth. Originally, Google's algorithms rapidly sorted through search terms, presenting results based on estimates of a site's relative importance (claimed to be a reflection of intuitive human judgments of importance). While Google's founders may have had noble origins, Google has strayed far from the path of truth.

Google learns, every time we search for something, what we think, like, value, and desire. Google's searches then return sites that reflect what we already think, like, value, and desire. Here is a way to see how Google works. Before I did my search on "belief" and "God," Google already knew that I am a religious believer; Google then returned sites favorable to belief in God. But if Richard Dawkins, famed atheist, had googled "belief in God," he would have gotten an entirely different set of sites; sites would have been critical of religious belief.

Suppose you were to google "George W. Bush." If Google had previously learned that you are a political conservative, it would give you sites that proclaim W. a great President (a decider, say, or socially conservative—which it knows you are, too, and so knows you take that ascription as a virtue). But if Google had previously learned that you are politically liberal, it would provide you with websites critical of Bush (reminding you, for example, that he got us into an unjust and unfunded war). In short, a Google search reflects your own shiny image back at you—it gives you sites that confirm your deeply held beliefs (in God, say, or in Bush), all the while shielding you from information opposed to your deeply held beliefs (in favor of atheism, say, or Marxism).

Why would Google do that? I suspect that Google does not want to disconcert us—it wants us to spend a lot of time perusing its search results. And, frankly, I don't think Google wants us to carefully peruse those pages because it cares about our dispassionate search for truth. Google's dastardly secret: Google wants us to buy things. And the more we look, the more likely we are to click on a link that takes us to a page where we will buy something (with a portion of those profits returned to Google). Google uses what it has learned about us to sell us something (by getting us to buy things from its advertisers). Check out the advertising links connected with your search results—see how curiously attuned they are to your own beliefs and values. Click on any search result, and check out how the ads on its pages are likewise curiously attuned to your own beliefs and values.

Curiouser and curiouser, Google favors pages that entice us to purchase cell phones, computers, and related accessories (which, in turn, benefit massive search engines); it is a vicious, non-truth-concerned circle. Each advertisement and sales pitch is precisely attuned to you—based on what it knows about what you think, like, value, and desire.

Google searches return sites that are fine-tuned to our own cherished beliefs and values because we don't, at bottom, like to have our most cherished beliefs and values challenged; we want our beliefs confirmed, celebrated, and praised (not denied or denigrated). If you like George Bush, you'll hastily and angrily close a webpage that contains a series of links blaming him for the Iraq war, the mortgage crisis, and increasing the deficit. And if you love God, you are likely to quickly shut down a page of links to Richard Dawkins, religion as the source of violence, and belief in God as infantile. Google taps into our psyches: our natural human tendency to prize evidence in favor of and to ignore and reject evidence opposed to our own beliefs. By finding what we want to hear, by "tickling our ears," Google ensures that we don't avert our gaze from the ads that entice us to buy things that they already know we like.

Here is a test. Go to Amazon and search "rototiller." Next, do a Google search on, say, Barack Obama. Carefully peruse the search results. Do the delivered sites support your like (or dislike) of Barack Obama? If the answer is yes, I have made half my case. Now look at the ads listed alongside the results (or on the sites Google has offered you). Any of them for rototillers? It may have taken Google a few days to learn of your Obama preferences, but it took only seconds for it to learn of your newfound interest in rototillers. Then it combined what it learned—about you and Obama and rototillers—into a uniquely organized and enticing page of search results and advertisements. How can you resist?

For all its virtues, Google is more likely to confirm what we already believe than to provide neutral but disagreeable and disconcerting information, at least about our most deeply held political, moral, and religious beliefs. It confirms our previously held beliefs for one simple reason—so that it can sell us something.

Yet, for all its faults, Google is astoundingly resourceful and useful, especially if we are aware of and attend to its flaws. I have belabored its flaws to make a point: it won't be very helpful as a guide to moral, political, and spiritual truth; it simply leads one to sites that confirm and away from sites that disconfirm one's most fundamental and even cherished beliefs. And yet it can inform us, with up-to-the-second news, about Tunisia, the latest presidential poll results, and the number of Muslims in Urumqi (and how the Chinese government is cracking down on them). It has videos of Bill Nye's debate with Ken Ham on evolution versus creation and the last known photo of Adolf Hitler. And where else could you learn before his mother did that Ben Affleck was splitting from Jennifer Garner? It is a veritable cornucopia of mundane and not so mundane information. Not so good, though, on belief in God. While Google can certainly help us identify the best mid-sized car and learn how to purchase it at the lowest price, it is not a reliable guide to the ultimate nature of reality.

Googling the Mind

Unfiltered, the mind would take in vastly more perceptual stimuli than we have the cognitive tools to handle. We are so bombarded with stimuli that we would go into instant cognitive overload if our minds didn't come preset with various governers and filters that limit and organize the mind's experiential input. Research in cognitive science has produced considerable evidence that human minds aren't best characterized as simple, undifferentiated general processors with a few basic faculties such as "memory," "perception," and "reason." Rather, in addition to these general activities, human minds also engage in a huge number of nonconscious conceptual activities that automatically and noninferentially generate beliefs and values to solve problems rapidly in particular domains of thought—possibly as adaptive mechanisms in response to selective pressures.

Take a break from reading, look up from this page, and notice the color and texture of the walls around you, the sounds of the music playing in the background, the gritty feel of the cover of the book (or the smooth feel of your computer), and the aroma of your coffee. While you have just attended more carefully to some sensations in your environment, note what you have not attended to but what is there: the color and texture of the ceiling and floor, the number of bricks in the wall, the temperature of the air, the look of any people in the room, the sounds of passing vehicles, and the blowing of the breeze (among many more things).

When you were reading, without any conscious decision whatsoever, your mind simply shut out the cacophony of sensations of color and sound and smell and taste and feel, allowing you to focus on your reading. If our minds were not so astoundingly good at ignoring most sensory stimulation, we would be so bombarded with information that we would go crazy. Our stingy minds, like Google, then, come outfitted with cognitive tools that allow in or attend to precious little stimuli.

Another set of cognitive tools organizes that information. These tools, like Google, accord with intuitive human judgments of importance (likely with primordial roots in survival). For example, if one is hungry, one is likely to be sensitive to food-source stimuli (smells, say) and to respond in terms of food acquisition (of the hunting/gathering/running-to-the-corner-market variety). If one is sufficiently cold, one will attend to one's discomfort and process information related to raising one's body temperature (building or entering a shelter, say, or making or putting on clothing). You can fill in the story for fleeing or fighting enemies, avoiding predators, and pursuing mates. Our minds are triggered by these evolutionarily important survival mechanisms, which in turn use various hardwired organizational tools to maximize survival strategies—getting food, securing a mate, avoiding a predator, or fending off an enemy. Although our native fight-or-flight response is set to a hair trigger, various cognitive tools are needed to plan the best escape route or select the best weapon. The initial cognitive tools that make rapid assessments for us ("You are in a dangerous situation," "You are hungry," "You are cold") don't produce judgments that we consciously decide or are taught; they are hardwired into the mind. Moreover, we process this information even further with additional hardwired cognitive information processors.

We are predisposed, for example, to prereflective "folk" physics. Folk physics is the way we naturally organize and understand, without any training whatsoever, the natural world. We find evidence of folk physics even in infants. Research in folk (sometimes called "naïve") physics has shown that within the first five months of life infants already expect that physical objects (1) tend to move only when launched through contact, (2) continue on inertial paths if not obstructed, (3) don't pass through other solid objects, (4) must move continuously through space (instead of teleporting from here to there), and (5) cohere as a bounded whole (unlike a cloud, a flame, or a pile of leaves).[1] An early emerging, natural cognitive faculty nonreflectively delivers commitments concerning the properties and motion of physical objects.

As humans mature, we likewise develop cognitive faculties that form and organize beliefs, independent of arguments or training, about living things (folk biology), social relations (folk sociology), mental activities (folk psychology), and hazard precautions (concerning contaminant-avoidance and other environmental dangers).[2] We have a hardwired fear of snakes, disgust for body sores, and suspicion of out-group persons (possible enemies); we thus avoid poisonous snake-bites, diseases spread through contamination, and death in war. We have hardwired cognitive tools devoted to the attraction of possible mates, acquisition of edible foods, and identification of friends, all of which assist human flourishing at many levels. It is not hard to see why, evolutionarily speaking, speedy judgments (without taking the time to consider arguments and counterarguments) on such matters might have been hardwired into the human psyche.

Although our cognitive faculties are tremendously useful (especially in survival situations), some of the folk judgments they produce are false. Folk biology, for example, encourages the belief that everything happens for a purpose, yet Darwinian biology rejects the view that camels developed humps for the purpose of traveling through deserts or that giraffes developed long necks for the purpose of reaching leaves higher on trees. Folk psychology has led some humans (many, even) to attribute intentions to rivers, say, and clouds. It is, without doubt, better for humans to judge in terms of purpose and intentions (and to do so instantly and prereflectively); by attributing purposes and intentions to things/people, we can plan our responses accordingly. If we think the lion wants to eat us (rather than run away), we can decide to climb a tree. If we decide that the approaching person has friendly intentions, we can strike up a conversation rather than strike him on the back of the head with a rock. Folk science allowed for the effortless sorting of information in ways of fundamental importance to human survival. But the quick, easy, and effortless is, sometimes, false.

We can learn a great deal about our cognitive faculties by understanding how they get things right. We can also learn a great deal about them by understanding how they get things wrong.

Let us think about some ways in which our minds, like Google searches, are sometimes capable of forming false beliefs. Like Google, our minds are trying to sell us something. We all prefer our own beliefs simply because they are our own; even more, we think we are rationally superior to people who disagree with us (even if we have no better evidence in support of our beliefs than our interlocutor does for theirs).

Think of your own political views. Suppose you are a liberal Democrat; do you really think that conservative Republicans are equally rational in their views? If you are an atheist, do you truly believe that religious believers are equally rational? Very likely, in both cases, you think yourself smarter (superior, even). As we consider those who disagree with us on important issues, a kind of native arrogance and smugness unwittingly emerges (this smugness emerges particularly when we are in faith- or politics-based groups and we laugh together, condescendingly, about those who reject our jointly held beliefs and practices). We think that those who disagree with us aren't just mistaken; they are irrational and even immoral.

"Irrational and immoral," you scoff. "Surely you're exaggerating?"

Consider one's assessment of a US presidential debate. Republicans think their candidate won handily and Democrats think the opposite. Both think that the other "does not get it" (does not REALLY understand). If they got it (REALLY understood), then they would share your belief. There is some sort of deliberate cognitive defect on their part, you think, one that prevents them from seeing what is so obvious (the intelligence, successes, and moral values of your candidate, say, and the stupidity, failures, and immorality of theirs). Exasperated, you shout, "If you would just listen . . ." You think they aren't just wrong; they are irrational. How about immoral? Political disagreements tap deep into the wellsprings of one's moral values. Republicans, on the one hand, favor individual autonomy and personal responsibility, while Democrats, on the other hand, favor interdependence and corporate justice. What looked, on the surface, to be an argument based on dispassionate reasons betrayed some of our most deeply held and even felt moral values. To reject the Republican (or Democrat) way, then, is perforce to reject one's deeply held and profoundly felt human values. And so the person who rejects your view is not, according to you, merely mistaken; she is irrational and even immoral.[3]

What our minds sell us, in cases like these, is a kind of unreflective and unwarranted self-satisfaction or self-confidence—false confidence, one might say. Minds also sell us a deep and righteous sense of being in-group (with a corresponding aversion to those who are out-group). A robust self-confidence is likely to promote success on the hunt or to make us more appealing to potential mates, while a deep and righteous sense of who is in one's group is likely to be helpful in dustups with out-groups (enemies, that is, competitors for scarce resources). Groups that rally their members around a cause or belief will be more cohesive and determined in fights with groups that lack deep and shared commitments. But, just as we cannot all be above average, we cannot all be right (not all group-rallying beliefs are true). Self-confidence and self-righteousness may win dates and fights, but they are infertile soil for truth.

And yet, for all its faults, our Google-mind is astoundingly resourceful, especially if we are aware of and attend to its flaws. Human cognition has split the atom, sent people to the moon, and cultivated wheat. Human beings have brewed crafty beers (after learning about yeast); built skycrapers that, well, scrape the sky (the Burj Khalifa in Dubai is over half a mile high);

and traveled to the bottom of the world's deep and astonishingly dark seas. We can see and even feel another's pain and then relieve it with altruistic sharing, robotic surgery, or psychotropic drugs. We have inferred a quantum world and thought our way back into the first mini-seconds of our universe nearly 14 billion years ago. We have learned that all people are created equal and thus that slavery is morally abhorrent. And we know many of the billions of people who live in our world, where and how they live, what they love, and how they are like and unlike ourselves (Google has helped with that). Of course, we also know many more quotidian things—what we ate for breakfast, the color of our house, and the names of our parents and next-door neighbors. We know that the sun will rise tomorrow and that China was united by Emperor Qin in 221 BCE (you can check the date on Google). I could go on and on, but you get the point: we know a lot. And we know a lot through the use of our cognitive faculties (fallible though they are).

Some of these true beliefs are acquired through the perfectly normal, natural, and unreflective use of such cognitive faculties as memory, sight, and theory of mind. Other true beliefs, especially scientific beliefs, require a highly disciplined and even chastened use of our cognitive faculties (relying on sophisticated forms of inference, abstract thinking, and high-level mathematics). Scientific beliefs are often deeply counterintuitive and require the rejection of the immediate deliveries of other cognitive faculties. For example, through sight we might believe that there is a rainbow out there—in the sky—but scientific theories of light and color tell us that rainbows are in here, in one's mind. Common sense tells us that a table is one, solid, and stable, while science tells us that the table is really made up of countless particles (or waves, or wave-particles) and mostly empty space (which combine to create the appearance of unity, solidity, and stability). And while it feels like we aren't moving (imagine what it feels like when you are in a convertible with its top down, going 80 mph), science tells us that we are on a planet hurtling through space at 483,000 mph (hold on to your hat!).

As we consider moral, political, philosophical, and spiritual truths, can we expect our cognitive faculties to succeed where Google has failed? The mind seems Google-like in at least this fundamental sense: reality includes too much information, too much stimulation, too many experiences for our minds to handle, and our cognitive faculties seem inadequate for culling and filtering this mass of information (at least when it comes to moral, political, philosophical, and spiritual truth). Given the rather dusty, evolution-shaped cognitive faculties that we have, maybe we have too slim a glimpse of reality to hope for moral, political, philosophical, and spiritual consensus. The wide variety of beliefs on free will, say, and the afterlife suggests a healthy caution, if not skepticism, about our ability to resolve deep philosophical disputes. Finally and sadly, like Google we have a bias in favor of confirming evidence and against disconfirming evidence for our most fundamental and cherished beliefs.

Googling Belief/Unbelief

Our Google-like cognitive faculties are highly selective gatekeepers and filters of information and experience. Some of our cognitive gatekeepers and filters—the agency-detecting device (ADD) and theory of mind (ToM)—inform and shape belief in God. According to Daniel Dennett and Richard Dawkins, God is nothing but a phantom in our minds, no more real than unicorns, dragons, or goblins.

Our Google-like cognitive faculties inform and shape unbelief as well. None of us is immune from various cognitive biases affecting our most cherished beliefs. Is atheism likewise a delusion?

Evolution, Dennett is fond of saying, is a universal acid: he claims that it eats through, undermines, our most cherished beliefs, including beliefs about God, value, meaning, purpose, culture, morality, and free will.[4] Evolution, he says, shows that our lives don't have a purpose, that we don't have free will, and that morality is an illusion fobbed off on us by our genes. And he devoted an entire book to arguing that evolution shows that God, like morality, is an illusion.

But if evolution is a *universal* acid, it should eat away the foundations of atheism as well as those of theism. Universal acids don't discriminate: evolution is equal opportunity—it can voraciously "explain away" every cultural form.

Atheism is not immune to the kinds of undermining explanations to which theism is allegedly susceptible.

You Just Believe That Because

When one learns of the natural cognitive mechanisms—the governers and filters—implicated in the production and sustenance of belief (usually of another's beliefs, not one's own), a certain sort of "you just believe that because . . ." psychologizing occurs. One firmly asserts that *you* believe *that* (fill in the blank), not because it is true or because you have good evidence for it, but only because of some psychological or moral defect. "You just believe morality is relative because you want to have sex with anyone you please." "You just bought that expensive car because you want people to think you are cool (or rich)." "You are a Republican because you just care about yourself."

"You just believe in God because your ADD and ToM went haywire."

But if psychological urges move religious folks, they just as likely move unreligious folks. In a mud fight everyone gets dirty. Just as the brain is implicated in belief, it is also implicated in unbelief.

"You are an atheist just because you are autistic."

The theist and atheist alike latch onto the other's perceived psychic breaks with reality.

As noted, we have a decided tendency to think that those who disagree with us are irrational or worse. "I am stubborn; you are pigheaded." Never thinking ourselves irrational, we denigrate those who disagree: we think that people who disagree with us irrationally acquired their beliefs in defiance of the evidence by acceding to their psychic urges. They are, in short, crazy. Consider some not-so-untypical political opinions: "Obama is not just mistaken; he is a Muslim Marxist who aims at nothing less than the total destruction of the United States of America as we know it." "Republicans aren't just wrong about human-caused global warming and evolution; they are anti-science, religiously fundamentalist nuts."

You just believe that because . . .

Rationality is the hammer we use to smash those who disagree with us.

Consider what happens when two cars arrive roughly simultaneously at a four-way stop and need to cross one another's road. In the United States, one should proceed in the order of arrival. Roughly simultaneous arrivals sometimes create a game of chicken. One person nudges out, hoping the other car recognizes his superior position with respect to the law. Sometimes, though, the other person, thinking she was first, also nudges ahead, asserting her superiority with respect to the law. A stubborn standoff ensues. Who will give in? I suspect there are many psychological biases engaged in these situations. (I will state them in the first person.) I am more likely to think that I got there first, simply by virtue of being me.[5] I am more likely to stubbornly dig in my heels if the other person asserts herself against me (and the more aggressively the other person asserts herself, the more defiantly I will maintain my belief that I got there first). Because of my firm conviction that I got there first, reinforced and even strengthened by the other's disagreement (but not bolstered by a single shred of new evidence), I will insist on the other's respect for my rights. Finally, even if I give in, I will think the other person selfish, irrational, and, if I get angry enough at her aggressive violation of my rights, crazy. Be glad I am not packing!

Atheists and theists can be like that. Neither of us is in an especially privileged position with respect to knowing the truth about the ultimate nature of reality. There are good reasons for and against belief in God. Atheists and theists reach their intellectual crossroads from different directions and face each other across the intellectual divide, often playing a game of intellectual chicken. A stubborn standoff ensues. Who will give in?

We have seen that there are many psychological biases engaged in these situations. (Again, I will state them in the first person.) Even setting aside the big guns—"irrationality mechanisms" for the theist and autism for the atheist—there are very ordinary and less spectacular biases engaged. I am more likely to think that I am right, simply by virtue of being me. I am more likely to stubbornly dig in my heels if the other person criticizes my views (the more aggressively the other person criticizes my views, the more confident I become in my belief). Because of my firm conviction that I am right, reinforced and even strengthened by the other's criticism (but not bolstered by a single shred of new evidence), I will insist on the other's respect for my beliefs. Finally, and here is a key difference, I won't give in. I will think the other person is irrational and, if I get angry enough at his aggressive criticism of my precious beliefs, crazy. Be glad I am not packing!

If we want the truth, psychopathologizing those who disagree with us, all the while valorizing our own allegedly dispassionate rational commitments, is likely to prove counterproductive. To be sure, religious belief is normal and natural. Does that make atheists abnormal and atheism unnatural? And atheism is very likely associated with inferential thinking. Does that mean that atheists are smarter or more in touch with reality than theists? What really follows from any of these suggestive studies? Just as there are psychic, cultural, and non-alethic impulses involved in religious belief, so, too, there are psychic, cultural, and non-alethic impulses involved in unbelief. To deny this would be folly. But what follows?

Are we all—believer and unbeliever alike—crazy or irrational?

Googling Humility

While Google has access to lots and lots of information, think of all the information that has not been digitized and located somewhere on the WWW. Google does not know how long I slept last night, what my mom sang to the infant me to help me fall asleep, how many hairs I have on my head, and countless more things about me. And Google does not know countless things about you (and every other human being). It does not know countless things about virtually every animal in the animal world, and every inanimate object in the non-living world as well. Google does not know whether there is an even or odd number of stars or the number of the blades of grass in my backyard. And countless things besides—well into the distant realms of space/time and into the tiniest micro-worlds (and past, present, and future).

Google has access to a very impressive amount of data, but what Google has access to is just a tiny fraction of the massive amount of information reality contains. Google is limited to delivering to us what some human beings somewhere have seen fit to digitize. Beyond that, a Google search has no access.

Reality is vastly bigger than the bits of information available to Google (or what can be googled). When Google gives us information, then, it has not searched the Whole of Reality. Impressive as the results of a Google search are, it is severely limited by its very nature to presenting us with a tiny fraction of a tiny digitized fraction of an infinitely massive Reality. We should not confuse a thorough Google search with an exhaustive search of Reality.

A Google search, then, is not the best way of settling matters on the Ultimate Nature of Reality.

Like Google, humans have access to just a tiny fraction of the information that constitutes reality. Even though we are capable of judging correctly that this wall is yellow, the visible light spectrum constitutes just .0035 percent of the entire electromagnetic spectrum. And while we know that the electromagnetic spectrum includes, in addition to the visible spectrum, radio waves, microwaves, infrared waves, ultraviolet waves, X-rays, and gamma rays, there may be electromagnetic waves we couldn't possibly be aware of. And while we see the world in the colors of the visible spectrum (ROYGBIV and their combos, say), we don't know what the world would look like in gamma or infrared waves (visible to a creature with a different set of cognitive faculties). Butterflies, for example, identify mates through the perception of ultraviolet markers, and reindeer use ultraviolet light to spot edible lichen. How butterflies and deer see the world is lost on us.

Likewise, we can hear less than 1 percent of the acoustic spectrum. While we know that bats use echolocation (sound waves and echoes) to "see" with sound, we have no idea what the bat "sees" (better, how bat consciousness differs from human consciousness). Are there acoustic frequencies beyond what we have measured (most of which we have not and could not hear)? Finally, while we have some sense of the sort of matter/energy that the world contains, we don't know anything about the hypothesized dark matter and dark energy that jointly make up 95 percent of the stuff of the universe.

We have peered (inferentially speaking) into the tiniest subatomic realms and have postulated, most famously, protons, neutrons, and electrons (and by "we" I mean some brilliant member of the human species but not exactly me). Further experiments seem to confirm the existence of even smaller particles, such as quarks and leptons. As I am writing, I read that the large hadron supercollider discovered a new subatomic particle today called the pentaquark.

Physicists have projected back in time to more than 13.7 billion years ago and have a reasonably good sense of the beginnings of our cosmos from 10^{-43} seconds. However, we have no idea what happened between 0 and 10^{-43} seconds. Moreover, we don't know what happened before t = 0 or what that even means.

But, and here is my point, is it not possible that there is more to subatomic reality than humans could ever possibly know? And is it not possible that something super-significant, mind-bogglingly, worldview-changingly awesome and surprising occurred between 0 and 10^{-43} seconds? And is it not possible that there is a before and maybe even a now or a bunch of nows (parallel universes?) that vastly exceed anything to which our puny but powerful cognitive equipment could possibly gain access? After all, our cognitive equipment was shaped and formed in our primate ancestors in response to compelling but banal survival and environmental pressures (to catch and then chew meat, say, not to discover $E = mc^2$); it seems like we might be better at mating (and we aren't so good at that) than at discovering the Nature of Ultimate Reality.

Then there are all those things of which science knows nothing: gods (if they should exist), and free will, and moral and aesthetic goodness and badness, and consciousness. Science is super, but, like Google, it is not Everything. Through Google we can gain access to a lot of information but never to the Whole of Reality. And physics, as successful as it is, can never gain access to the Whole of Reality (maybe not even to the whole of the natural world; who knows?).

And so, given our tiny cognitive equipment and our thin grasp of the Whole of Reality, mightn't we expect some disagreement about the Ultimate Nature of Reality? The appropriate response to the disproportion between our cognitive finitude and Reality's astounding infinitude is not the intellectual arrogance to which we are so tempted, but *intellectual humility*.

When made aware of the disproportion between Infinite Reality, on the one hand, and our finite cognitive equipment, on the other, we should say, to all of our fellow human inquirers who are doing their best to figure it all out but who have come to decidedly different views than our own, "I respect you and honor your serious and brave attempt to grasp the Nature of Reality and to live your life faithfully in accord with your understanding of Reality. Help me understand your beliefs and, if there is any time left, I would be happy to share mine." And even if there is not time to explain your own beliefs, before you leave, doff your hat, bow deeply, and express your gratitude to them for so poignantly sharing their struggle to understand. Then go home and ponder what you have learned from them (and never diminish them by thinking them your inferior either rationally or morally, and never think yourself their intellectual or moral superior).

Bodies, Minds, and Gods

It is simply remarkable that humans have been able to theorize so successfully about the empirical world given our modest cognitive origins. Our cognitive faculties evolved from primate ancestors in response to mundane selective pressures to, say, eat and mate. As early *Homo sapiens* groups increased in size, the need to understand other members of the group and to communicate that understanding may have led to increased abilities to think and to speak in words. Somewhere along the way, humans learned the very useful practice of attaching numbers to various things and developed the ability to think about *what if?* (that is, we learned to think counterfactually). Increased thinking capacity and abilities combined with numerical thinking and counterfactual thinking to produce, very late in human history, chemistry, biology, anatomy, and, ultimately, physics. Somehow humans leaped, 200,000 years after *Homo sapiens* first appeared, from 1, and 2, and 3 to $E = mc^2$.

Those very ordinary cognitive capacities likewise shaped religious beliefs (and their denials) and practices. Our ability to function well in increasingly large groups required us to understand human agency at a deeper level. More importantly, successful large group interactions required an ability to grasp very quickly what others were thinking or feeling. We humans are a suspicious lot, you see, and we need to know whom we can trust and whom we cannot (and very quickly). Our ability to read minds—other people's thoughts, feelings, and desires—was absolutely essential to human survival. Thus there emerged a highly attuned ToM, one that would, in conjunction with our agency-detecting device, shape God-beliefs (along with other perfectly ordinary cognitive faculties such as promiscuous teleology). We think God with the very same cognitive tools we think people and, sometimes, porpoises; nothing special here.

And if we think God with our very ordinary cognitive faculties, we unthink God with our very ordinary, mundane cognitive faculties as well. And if we have cognitive biases that incline us to belief, we very likely have cognitive biases that incline us to unbelief.

Realizing that we are, at bottom, created from evolutionary dust is terrifying. What makes us good at thinking about avoiding predators and attracting mates makes us not so good at thinking about physics and even less good at thinking about philosophy (God, freedom, and immortality). Our dusty cognitive equipment comes with a stern warning: don't exceed the limits. Have we bumped up against our cognitive limits already, before we even get to gods?

Theists, it seems to me, should resist at least some skeptical worries when thinking about the humble origins of our embodied minds. Theists believe that we are God's creatures and that everything God created is good. If so, our bodies are good and embodied existence is good. And, whatever one thinks of the soul-body distinction, human consciousness is intimately intertwined with our brains. We think—not our brains, not our souls; but when we think, our brains are involved (and so are our bodies and our communities). If God speaks to us or loves us, then we process God's speech and love with our brains. If we have faith, then there is a substratum in our brains that processes and sustains our belief in God. Understanding the brain, then, is one way for us to come to a better understanding of our faith. If God created brain-based consciousness, then understanding brain-based consciousness (even of God himself) is good (and so the sciences of the brain are not to be feared).

Theists likewise have (or should have) a robust sense that humans aren't gods. We aren't omniscient, and we don't have a godlike, timeless/spaceless perspective on All of Reality. We are, first and foremost, creatures (with all that that entails). We are located at this point in human history on this tiny planet located within this otherwise apparently insignificant galaxy in a far-off corner of the universe. We have the cognitive faculties of humans—above apes but below angels (and well below God). When we combine our radical contingency with our intellectual finitude, it is no wonder, to borrow from St. Paul, that "we see through a glass darkly."

We have landed in a neighborhood familiar to believer and unbeliever alike: we aren't gods. Not being gods, we are creatures, and therefore creaturely, finite, temporal modes of fallible understanding are appropriate. None of us, believer and unbeliever alike, has a godlike perspective on any of these issues; we cannot see inside another's mind to see either their beliefs or their mental shortcomings (and we aren't all that honest about our own beliefs and shortcomings). Humility, then, not arrogant pronouncements on another's character or beliefs, seems the order of the day. As we reflect on the rationality of belief and unbelief, a little humor, a lot of charity, and even more humility seem called for.

Inference, Intuition, and Rationality

Rationality, as used throughout this book, involves doing the best one can to use one's cognitive faculties to discover the truth or get in touch with reality; doing one's best involves the use of both intuitive and inferential faculties. However, discussions of rational belief typically overvalue inferential thinking and correspondingly denigrate intuitive thinking. Richard Dawkins, for example, asserts: "As a lover of truth, I am suspicious of strongly held beliefs that are unsupported by evidence."[1] Psychological studies of intuitive versus inferential thinking typically overvalue inferential reasoning.[2] And philosophers, who trade in arguments, routinely value inferential over immediately acquired beliefs.

I will argue that inferential thinkers (correlated with atheism), at least on philosophical matters (including belief in God), are no more rational than intuitive thinkers (correlated with theism). Another, more polemical, way to put it: an atheist who has cultivated unbelief on the basis of an argument (i.e., inferentially) is not thereby more rational than a theist who believes simply on the basis of her God-faculty (i.e., intuitively).

We finite and fallible human believers cannot acquire true beliefs about the world without a vast set of intuitive or basic beliefs. For example, we intuitively and rationally believe that $2 + 2 = 4$, that causing the suffering of innocents is wrong, that the future will be like the past, and that there is a world outside our minds. Intuitive thinking produces pervasive, foundational, and even true beliefs; we simply cannot believe much of anything without such foundational beliefs. Very likely, *most* of our deepest and most pervasive beliefs are intuitive.[3] Such beliefs are *prima facie* rational (assuming, as we charitably and sympathetically should, that one is doing one's best to acquire the truth). Of course, people have also intuitively believed that some people are naturally suited to be slaves, that the earth is flat, and that women aren't rational. Intuitive beliefs, though often rational, aren't infallible.

But not all of one's beliefs are or should be held intuitively. Some of one's rational beliefs are inferential. A jury should judge the guilt of a defendant only after careful consideration of the evidence pro and con; a person might decide to buy a certain shampoo after reading several reviews; a biologist should slowly come to believe in descent with modification after a careful reading of the *Origin of Species*; and Einstein should affirm that E = mc² only after some thought experiments and working out the mathematics. Of course, most of us (even most scientists) don't know the evidence supporting E = mc², and so our rational belief that E = mc² is based not on scientific evidence but on testimony.

Rational beliefs, then, come in two forms—intuitive and inferential. If we were to restrict ourselves to inferential beliefs, we would have nothing to believe. If we have rational inferential beliefs, we must also have rational noninferential beliefs.[4] So reason does not, cannot, and should not liberate us from intuitive beliefs.

At the basis of every inference, we find assumed intuitions. Reliance on intuition is often disguised by the remarkable complexity of philosophical arguments. Beneath the symbols, modalities, and nested propositions, one finds an intuition. In every philosophical argument, there is at least one fundamental premise that cannot be argued for. Dig deep enough, and one will find the unargued starting point. This unargued starting point is an intuition—an immediate, noninferential judgment. Such intuitions may be elicited by stories, motivated by cases, critiqued by counterexamples, or appealed to in theories, but they aren't and cannot be argued for. One "gets them" (or not).[5]

Inference itself requires assumptions about the nature and power of logic, for example, or the applicability and meaning of probability. Inference likewise assumes how reality seems to us (intuition) in every domain of human inquiry. Our ordinary, commonsense beliefs rely on intuitions (noninferential assumptions) about space and time, the reliability of sense perception, belief in the past, and belief in an external world. Scientific inferences assume without argument the uniformity of nature, the inductive principle, and truths of mathematics.

Although we must rely on our intuitions, we aren't so metaphysically astute that we can clearly and certainly perceive those involved in, for example, an argument for (or against) the existence of God, for an absolute and universal moral standard, or for the immateriality of reality. Relevant intuitions in these fields might include claims that an infinite regress of causes is absurd, that moral statements require grounding, and that sensory appearances can be adequately accounted for without reference to a material world. Discussions in political theory, ethics, the meaning of life, the nature of human persons, determinism, and free will rely on crucial premises that are intuitive. Every substantive philosophical belief about reality, then, betrays one's commitment to fundamental, intuitive beliefs.

Now on to belief in God. Everyone's belief or disbelief in God, inferential or not, is grounded, ultimately, in intuition.[6] For most religious believers, belief in God is intuitive—that is, nonreflective or noninferential. Those whose belief in God is inferential rely on arguments that are grounded in intuitions (the principle of sufficient reason, for example, or the objectivity of morality).[7] I suspect that many, perhaps most, unbelievers are atheists not through careful assessment of theistic arguments: they are, instead, conformity or prestige biased, maybe autistic (who knows?). What about those whose rejection of theism is consciously inferential? Even those atheists, I am arguing, had to rely on intuitive (noninferential) epistemic principles (perhaps assuming that belief in God must be accepted or rejected according to the canons of scientific rationality) or metaphysical principles (perhaps rejecting the principle of sufficient reason or the objectivity of morality).

We can gain some understanding of the role of intuition in the formation of philosophical beliefs by placing Plato and Aristotle side by side. Plato was deeply suspicious of sense perception, hoping to escape from this elusive and illusive shadowy world into the Real and the Good. Although deeply influenced by his teacher, Aristotle was constitutionally disposed to muck about, relish, and find reality in the very material world that Plato despised. Aristotle's philosophy affirms this world, particulars, and matter. While both argued for their particular worldviews, they relied fundamentally on different intuitions. Both could account equally well for all that humans experience. And yet their conclusions were driven by their differing intuitions that the truth lies in this direction rather than that one. While their intuitions sometimes found expression in arguments, *intuition, not inference, drove the development of their worldviews.*

Philosophical thinking is deeply and irremediably grounded in intuition. Philosopher Hilary Kornblith argues that there is no reason to think philosophical reflection (inference) is better than nonreflective (noninferential) thinking.[8] Empirical studies have shown that, due to confirmation bias and our tendency to rationalize (after the fact), reflection is often inaccurate. When challenged, reflection yields both rationalizations and a false sense that we have good grounds. Reflecting on our beliefs, then, doesn't always get us closer to the truth.

We seldom acquire beliefs as the result of coolly rational, explicit, and dispassionate attention to arguments (though we pride ourselves on having done so). Our beliefs and practices are more often the product of universally pervasive, sometimes unconscious (implicit) processes that are automatically activated in a wide variety of circumstances. "At the nexus of social psychology, cognitive psychology, and cognitive neuroscience has emerged a new science called 'implicit social cognition' (ISC). This field focuses on mental processes that affect social judgments but operate without conscious awareness."[9] These pervasive biases are triggered involuntarily and without one's awareness or intentional control. For example, while most Americans explicitly disavow racism, studies show that most Americans in fact have anti-black prejudices that move them to believe and act in various ways. And most Americans are biased with respect to age, gender, skin color, ethnicity, nationality, race, sexual orientation, class, weight, and, of course, religion. Such biases dispose some of us to beliefs and practices that dispossess and disadvantage black people or women or people who are overweight.[10] Even after sustained training, thorough self-examination, and genuine desire for change, these biases persist and surface in ways that harm those on the receiving end. These unconscious mental processes operate clandestinely, directing one's reasoning processes.

When confronted with an understanding of those unconscious mental processes or biases, I sometimes tell myself a story, one that makes sense of my beliefs and actions, a story that rationalizes my behavior (within which I am a careful, rational reflector and virtuous agent). Such stories, which come after but are offered as the explicit reasons for my beliefs or actions, are nothing more than confabulations (bullshit, to use the non-technical term). I make myself the rational hero of my own drama. And I feel better—more rational, more virtuous—after hearing and heeding my own story. Indeed, these stories make me more confident in my beliefs, more proud of my intellectual prowess, more assured of my virtue. And sometimes move me further from the truth.

Kornblith criticizes the philosopher's insistent demand for rational reflection for two reasons: (1) because of the necessity of intuitions for every philosophical argument; and (2) because of our very human tendency to offer rationalizations of our previously (intuitively) held beliefs. After providing such rationalizations, subjects are often more confident of their belief, but for no good reason. While they find their alleged justifications of their initial beliefs persuasive, they are simply bad reasons that offer no logical support for their initial beliefs. Just as humans are influenced in a wide variety of non-truth-conducive ways in their acquisition of intuitive or immediate beliefs, they are equally susceptible to non-truth-conducive ways of rationalizing their beliefs. Kornblith writes: "The idea, then, that by reflecting on the source of our beliefs, we may thereby subject them to some sort of proper screening, and thereby improve on the accuracy of the resulting beliefs, is simply misguided. When we reflect in this way, we get the impression that we are actually providing some sort of extra screening of our beliefs, and we thus have the very strong impression that we are actually doing something to insure that our beliefs are, indeed, reliably arrived at. But this is not what we are doing at all."[11]

This act of what Kornblith calls *self-congratulation* does little more than make us feel better about ourselves and superior to those we have judged defective. As Kornblith points out, we have a strong tendency *to prefer beliefs simply because they are ours.* And we have a strong tendency toward *belief conservatism*—to preserve or conserve our already-held beliefs. We have a tendency to notice and favor evidence that supports our previously held beliefs and to ignore or discount evidence that opposes them. We easily remember evidence in favor of our beliefs, while we just as easily forget evidence that opposes them. On those precious few occasions when we do stop and "reflect," little wonder if we find that our previously held beliefs are overconfidently held and asserted.

Since philosophical arguments essentially rely on intuitions, neither resting on intuition nor relying on argument is any better at gaining the truth. Indeed, perhaps both are equally bad at gaining philosophical knowledge. Unlike many other intuitive beliefs, with philosophical intuitions we cannot check the facts to see if they are reliable. We have no belief-independent access to the philosophical world.

With respect to philosophical matters (including belief or disbelief in God), then, intuition and inference are on epistemically equal ground. If rationality involves doing the best one can to get in touch with the truth, neither intuition nor inference has an epistemic advantage.

While scientists, psychologists, and philosophers value inferential thinking over intuitive thinking, and while inferential thinking may be associated with higher IQ, human beings cannot avoid reliance on intuition. The situation is all the more pressing in matters philosophical. Scratch an inferentialist and you will find an intuitionist. That is, look carefully at a philosopher's proffered argument, and you will find an essential, intuitively accepted premise. Even for the most ardent demander of evidence, argumentative reasoning starts with intuitions.[12]

Religious belief may be more nonreflective, but religious believers aren't evidence insensitive. And atheists may be more inferential, but arguments assume intuitions. Neither has an epistemic advantage.[13]

Notes

CHAPTER 1

1. Unless otherwise noted, Scripture quotations are taken from the New International Version.

2. Paul Bloom, "Is God an Accident?," *Atlantic Monthly*, December 2005.

3. Eben Alexander, *Proof of Heaven: A Neurosurgeon's Journey into the Afterlife* (New York: Simon and Schuster, 2012).

4. Alexander, *Proof of Heaven*, 45.

5. Alexander, *Proof of Heaven*, 71.

6. Max Read, "*Newsweek* Cover Story or Internet Posting about Drugs? A Quiz," *Gawker*, October 8, 2012, http://gawker.com/5949892/newsweek-cover-story-or-internet-posting-about -dugs-a-quiz.

7. Sam Harris, "This Must Be Heaven," *The Blog*, October 12, 2012, http://www.samharris.org/blog/item/this-must-be-heaven.

8. Victor Stenger, "Not Dead Experiences (NDEs)," *Huffington Post*, October 11, 2012, https://www.huffingtonpost.com/victor-stenger/not-dead-expereirnces-nde_b_1957920.html.

9. Oliver Sacks, "Seeing God in the Third Millennium," *Atlantic Monthly*, December 2012, https://www.theatlantic.com/health/archive/2012/12/seeing-god-in-the-third-millennium /266134/.

10. Maia Szalavitz, "Q&A: An Interview with Oliver Sacks," *Time: Health & Family*, October 27, 2010, http://healthland.time.com/2010/10/27/mind-reading-an-interview-with -oliver-sacks/. Until 2013, Asperger's syndrome was considered a separate condition, but since then it has been considered a form of autism, which can include, among many other things, difficulty in social interactions, inability to read and respond to normal social cues, and difficulty in making friends.

11. Jimo Borjigin et al., "Surge of Neurophysiological Coherence and Connectivity in the Dying Brain," *Proceedings of the National Academy of Sciences of the United States of America* 110.35 (2013): 14432–37.

12. Quoted in Robert Hercz, "The God Helmet," http://www.skeptic.ca/Persinger.htm.

13. Pehr Granqvist et al., "Sensed Presence and Mystical Experiences Are Predicted by Suggestibility, Not by the Application of Transcranial Weak Complex Magnetic Fields," *Neuroscience Letters* 379.1 (April 29, 2005): 1–6.

14. See an interview with Blackmore at https://www.youtube.com/watch?v=Zo-ac hedLMs.

15. Dean Hamer, *The God Gene: How Faith Is Hardwired into Our Genes* (New York: Doubleday, 2004).

16. "The God Gene," *Time* magazine, November 29, 2004.

17. Carl Zimmer, "Faith-Boosting Genes: A Search for the Genetic Basis of Spirituality," *Scientific American* 291.4 (October 2004).

CHAPTER 2

1. In what follows, I rely on the work of Scott Atran, *In Gods We Trust: The Evolutionary Landscape of Religion* (Oxford: Oxford University Press, 2002); Justin Barrett, *Why Would Anyone Believe in God?* (Lanham: AltaMira, 2004); Pascal Boyer, *The Naturalness of Religious Ideas: A Cognitive Theory of Religion* (Berkeley: University of California Press, 1994); Stewart Guthrie, *Faces in the Clouds* (Oxford: Oxford University Press, 1993); Robert N. McCauley, *Why Religion Is Natural and Science Is Not* (Oxford: Oxford University Press, 2011).

2. Simon Baron-Cohen (principal author), *Mind Reading: The Interactive Guide to Emotions* (London: Jessica Kingsley Publisher, 2003).

3. Barrett, *Why Would Anyone Believe in God?*, 36–37.

4. The ubiquity of beliefs in gods, much as beliefs in minds and the regularity of nature, is some preliminary evidence of a God-faculty. Anthropologist Atran, *In Gods We Trust*, 57, writes: "Supernatural agency is the most culturally recurrent, cognitively relevant, and evolutionarily compelling concept in religion. The concept of the supernatural is culturally derived from an innate cognitive schema." Atran's move from observing the recurrence of belief in gods to linking such beliefs to a natural part of human cognition is not grounded only on the commonness of belief in gods. Rather, Atran and other cognitive scientists of religion have begun identifying various cognitive systems that, working in concert, seem to give belief in gods intuitive support. Following anthropologist Stewart Guthrie, Atran argues for the importance of an agency-detection system that has evolved to detect predators, prey, and other people in the environment.

5. Guthrie, *Faces in the Clouds*; Atran, *In Gods We Trust*.

6. Stewart Guthrie, "A Cognitive Theory of Religion," *Current Anthropology* 21.2 (1980): 181–203; Guthrie, *Faces in the Clouds*.

7. I will set aside discussion of the exact nature of the God-faculty and consider instead some natural cognitive faculties that dispose us to religious belief. I will not discuss whether it is a single module of the mind-brain or is a complex involving various parts of the mind-brain. Pascal Boyer, for example, rejects the claim that there is a single module that produces religious beliefs: "The first thing to understand about religion is that it does not activate one particular capacity in the mind, a 'religious module' or system that would create the complex set of beliefs and norms we usually call religion. On the contrary, religious representations are sustained by a whole variety of different systems." Boyer, "Why Is Religion Natural?," *Skeptical Inquirer* 28.2 (March/April 2004).

8. Jesse M. Bering, "Intuitive Conceptions of Dead Agents' Minds: The Natural Foundations of Afterlife Beliefs as Phenomenological Boundary," *Journal of Cognition and Culture* 2.4 (2002): 263–308; Jesse M. Bering, "The Existential Theory of Mind," *Review of General Psychology* 6 (2002): 3–24.

9. Jesse M. Bering and Dominic Johnson, "'O Lord . . . You Perceive My Thoughts from

Afar': Recursiveness and the Evolution of Supernatural Agency," *Journal of Cognition and Culture* 5 (2005): 118–42.

10. Bering, "Intuitive Conceptions of Dead Agents' Minds"; Bering, "The Existential Theory of Mind."

11. Bering, "Intuitive Conceptions of Dead Agents' Minds"; Bering, "The Existential Theory of Mind."

12. Paul Bloom, *Descartes' Baby: How Child Development Explains What Makes Us Human* (London: William Heinemann, 2004).

13. Pascal Boyer, *Religion Explained: The Evolutionary Origins of Religious Thought* (New York: Basic Books, 2001).

14. Deborah Kelemen, "Are Children 'Intuitive Theists'? Reasoning about Purpose and Design in Nature," *Psychological Science* 15.5 (2004): 295–301.

15. Justin L. Barrett and Rebekah A. Richert, "Anthropomorphism or Preparedness? Exploring Children's God Concepts," *Review of Religious Research* 44.3 (2003): 300–312; Barrett, *Why Would Anyone Believe in God?*

16. For a more complete review of the relevant literature, see Barrett and Richert, "Anthropomorphism or Preparedness?"; and Barrett, *Why Would Anyone Believe in God?* (esp. chapter 6). Evidence is thinner for the intuitiveness of immortality, but the presumption that minds don't automatically stop or die at some point has been suggested by Jesse M. Bering, "The Folk Psychology of Souls," *Behavioral and Brain Sciences* 29 (2006): 453–98; and Bloom, *Descartes' Baby*.

17. Paul Bloom, "Religion Is Natural," *Developmental Science* 10 (2007): 147–51.

18. Barrett, *Why Would Anyone Believe in God?*, 124.

19. Boyer, *The Naturalness of Religious Ideas*.

20. Atran, *In Gods We Trust*, 4.

21. One might think that this shows that God is all in the mind (people like Richard Dawkins and Daniel Dennett do). We are, after all, relentless anthropomorphizers who seem unable to shut off ToM's spout when it starts spewing. We find people-like agents in clouds, stars, thunder, and even corpses. Because we are naturally inclined to believe that our loved ones survive death and pass on to a spiritual form of existence, dead bodies serve as containers for gods: sometimes burning the body releases the soul. We take a little bit of ADD, shake it together with a dash of ToM, and create gods. Voilà, all in the mind. We will return to this claim in chapter 4.

22. Joseph Bulbulia, "Religious Costs as Adaptations That Signal Altruistic Intention," *Evolution and Cognition* 10 (2004): 23.

23. Ara Norenzayan, *Big Gods: How Religion Transformed Cooperation and Conflict* (Princeton: Princeton University Press, 2013).

24. Ara Norenzayan, *Big Gods: How Religion Transformed Cooperation and Conflict* (Princeton: Princeton University Press, 2013), ch. 3.

25. Dominic Johnson and Oliver Krüger, "The Good of Wrath: Supernatural Punishment and the Evolution of Cooperation," *Political Theology* 5.2 (2004): 157–73.

26. Richard Sosis, "Why Aren't We All Hutterites? Costly Signaling Theory and Religious Behavior," *Human Nature* 14 (2003): 91–127. Perhaps religious rituals are more powerful than secular rituals because costly signaling by itself is not sufficient to weed out freeloaders. Religious rituals add supernatural punishment theory to the motivational mix.

CHAPTER 3

1. Richard Swinburne conceives of belief in God as *like* belief in a scientific theory and brings science-like modes of reason to bear on rational belief in God (again, worth noting: he does not think that belief in God *is* a scientific theory, and so belief in God is not, according to him, in competition with any particular scientific theory); see Richard Swinburne, *The Existence of God* (Oxford: Oxford University Press, 2004). If one is inclined to think of belief in God in this manner, Swinburne's work would repay one's efforts.

2. This view is most famously held by contemporary philosophers Alvin Plantinga and Nicholas Wolterstorff, who acknowledge their allegiance to the eighteenth-century Scottish philosopher Thomas Reid. Proponents who hold views in the neighborhood include William Alston, George Mavrodes, John Greco, Alvin Goldman, Ludwig Wittgenstein, and G. E. Moore.

3. There are, of course, versions of rationality that have little to do with the truth. For example, some hold that a person is rational insofar as she is successful in achieving her goals or purposes.

4. This is not to deny that some people are psychologically disposed, at least on some occasions, to avoid the truth.

5. The burden of proof, then, is on the theist, but "there is no evidence to favor the God Hypothesis" (Richard Dawkins, *The God Delusion* [Boston: Houghton Mifflin, 2006], 58–59). So the rational position, he concludes, is agnosticism shading off into atheism. Dawkins contends that rational belief in God requires the support of evidence; lacking such evidence, belief in God is irrational.

6. William K. Clifford, "The Ethics of Belief," *Contemporary Review* (1877); reprinted in William K. Clifford, *Lectures and Essays*, ed. Leslie Stephen and Frederick Pollock (London: Macmillan and Co., 1886).

7. Clifford, "The Ethics of Belief," 339.

8. Clifford, "The Ethics of Belief," 346 (my emphasis).

9. This set of evidence may seem unduly restrictive. You might think that one's perceptions can be corroborated by the perceptions of one's fellow human beings (so evidence should include what others tell us they perceive). However, the Enlightenment conception of rationality that I am representing here was committed to absolute certainty. I can be absolutely certain only of what I personally experience or rationally grasp. Since what others tell us is sometimes wrong, I cannot be certain of the testimony of others.

10. There are powerful defenses of evidentialism; see especially Earl Conee and Richard Feldman, "Evidentialism," *Philosophical Studies* 48 (1985): 15–34. For a good discussion of evidentialism, see the essays in Trent Dougherty, ed., *Evidentialism and Its Discontents* (Oxford: Oxford University Press, 2011).

11. In William G. Lycan, *Judgement and Justification* (Cambridge: Cambridge University Press, 1988).

12. The view of rationality most opposed to the Reidian version that I have defended is strong evidentialism. Strong evidentialism holds that a belief *p* is rational if *p* is supported by evidence, where evidence is taken either as beliefs one holds or as propositional arguments one has grasped. According to strong evidentialism, then, belief in God is rational if and only if it is supported by evidence, again usually taken to mean that one has acquired belief in God through reflection on and embrace of a sound propositional argument for the existence of God (such as the cosmological argument or the argument from design). The rationality or

irrationality of belief in God, according to strong evidentialism, is based entirely on theistic arguments and counterarguments (say, the problem of evil or the hiddenness of God). Although I reject strong evidentialism for the reasons listed above, I will show how cognitive science of religion (CSR) is relevant to strong evidentialism as we go along.

13. John Greco, *Putting Skeptics in Their Place: The Nature of Skeptical Arguments and Their Role in Philosophical Inquiry* (Cambridge: Cambridge University Press, 2000).

14. Raymond Nickerson, "Confirmation Bias: A Ubiquitous Phenomenon in Many Guises," *Review of General Psychology* 2 (1998): 175.

15. Defenses of the position—let us call it unfetchingly the Reid-Wolterstorff-Plantinga position—that I have outlined in this chapter include the following: Alvin Plantinga and Nicholas Wolterstorff, eds., *Faith and Rationality: Reason and Belief in God* (Notre Dame: University of Notre Dame Press, 1984); Thomas Reid, *An Inquiry into the Human Mind on the Principles of Common Sense* (1764), *Essays on the Intellectual Powers of Man* (1785), and *Essays on the Active Powers of the Human Mind* (1788); Nicholas Wolterstorff, *Reason within the Bounds of Religion*, 2nd ed. (Grand Rapids: Eerdmans, 1984), and *Thomas Reid and the Story of Epistemology* (Cambridge: Cambridge University Press, 2001); Alvin Plantinga, *Warrant: The Current Debate* (Oxford: Oxford University Press, 1993), *Warrant and Proper Function* (Oxford: Oxford University Press, 1993), and *Warranted Christian Belief* (Oxford: Oxford University Press, 2000); William Alston, *Perceiving God: The Epistemology of Religious Experience* (Ithaca, NY: Cornell University Press, 1991), and *The Reliability of Sense Perception* (Ithaca, NY: Cornell University Press, 1993); John Greco, *Putting Skeptics in Their Place*, and "How to Reid Moore," *Philosophical Quarterly* 52.209 (2002): 544–63; and Michael Bergman, *Justification without Awareness* (Oxford: Oxford University Press, 2006). There are countless other contemporary approaches to knowledge, rationality, truth, and warrant (and related subjects). The view outlined in this chapter, the Reid-Wolterstorff-Plantinga view, is viable, defensible, and suited to our nature as finite creatures. But it is just one of the many approaches that philosophers take to knowledge. For different accounts, see Michael Williams, *Problems of Knowledge: A Critical Introduction to Epistemology* (Oxford: Oxford University Press, 2001); Robert Audi, *Epistemology: A Contemporary Introduction to the Theory of Knowledge* (New York: Routledge, 2010); Fred Feldman, *Epistemology* (Upper Saddle River, NJ: Pearson, 2002); Richard Fumerton, *Epistemology* (Malden, MA: Wiley-Blackwell, 2006); Richard Swinburne, *Epistemic Justification* (Oxford: Oxford University Press, 2001); and Jay Wood, *Epistemology: Becoming Intellectually Virtuous* (Downers Grove, IL: InterVarsity, 1998).

CHAPTER 4

1. According to strong evidentialism, one's belief in God would be rational if and only if one had consciously grasped a sound argument, like the cosmological argument or the argument from design, for the existence of God.

2. A study suggests that nearly two-thirds of Americans claim to have had profound religious experiences. See Rodney Stark, *What Americans Really Believe* (Waco, TX: Baylor University Press, 2005).

3. Alvin Plantinga, *Warranted Christian Belief* (Oxford: Oxford University Press, 2000), 173–75.

4. Darwin himself, while rejecting the universality of belief in God, concedes the universality of religious beliefs: "The belief in God has often been advanced as not only the greatest

but the most complete of all the distinctions between man and the lower animals. It is however impossible to maintain that this belief is innate or instinctive in man. On the other hand a belief in all-pervading spiritual agencies seems to be universal, and apparently follows from a considerable advance in man's reason, and from a still greater advance in his faculties of imagination, curiosity and wonder" (Charles R. Darwin, *The Descent of Man* [London: J. Murray, 1874], 609–19).

5. Richard Dawkins, *The God Delusion* (Boston: Houghton Mifflin, 2006), 184.

6. Helen De Cruz has shown that arguments for God's/the gods' existence arise only under highly specific cultural conditions—that is, presence of a leisure class, atheism, or contact with competing religious frameworks. Therefore, we find arguments in ancient Greece, ancient India, etc. If arguments played an essential role in the production of religious beliefs, that would be puzzling, since it would not explain why people across cultures are religious believers. See Helen De Cruz, "Cognitive Science of Religion and the Study of Theological Concepts," *Topoi* 33.2 (2013): 487–97. Robert McCauley calls religious belief natural and natural theology cognitively unnatural. *Why Religion Is Natural and Science Is Not* (Oxford: Oxford University Press, 2011).

7. Georg Wilhelm Friedrich Hegel, *Encyclopaedia of the Philosophical Sciences*, 3rd ed., trans. William Wallace (1830).

8. According to strong evidentialism, one's immediately acquired belief in God would not be rational unless or until one supported it with a sound argument for the existence of God.

9. Pascal Boyer, *Religion Explained: The Evolutionary Origins of Religious Thought* (New York: Basic Books, 2001), 76.

10. The more scholarly versions of evolutionary debunking arguments share the claim that the relevant belief-producing faculties are either unreliable (they produce mostly false beliefs) or insensitive to truth (the ground of the truth of a belief does not produce the belief). Evolutionary debunking arguments are typically offered against rational moral or religious beliefs (and extended to, for example, belief in free will, belief in an enduring self, etc.). According to Guy Kahane, evolutionary debunking arguments share the following structure:

Causal premise: S's belief that p is explained by X.

Epistemic premise: X is an off-track process.

Therefore, S's belief that p is unjustified.

I aim at this sort of argument. For scholarly defenders of evolutionary debunking arguments against belief in God, see Paul E. Griffiths and John S. Wilkins, "Evolutionary Debunking Arguments in Three Domains: Fact, Value, and Religion," in *A New Science of Religion*, ed. Greg Dawes and James Maclaurin (New York: Routledge, 2012), 133–46; Guy Kahane, "Evolutionary Debunking Arguments," *Noûs* 45 (2011): 103–25. For scholarly objectors to evolutionary debunking arguments against belief in God, see Michael J. Murray and Andrew Goldberg, "Evolutionary Accounts of Religion: Explaining and Explaining," in *The Believing Primate: Scientific Philosophical and Theological Reflections on the Origin of Religion*, ed. Michael J. Murray and Jeffrey Schloss (Oxford: Oxford University Press, 2009), 179–99; Joshua Thurow, "Does Cognitive Science Show Belief in God to Be Irrational? The Epistemic Consequences of the Cognitive Science of Religion," *International Journal for Philosophy of Religion* 74 (2013): 77–98; Jonathan Jong, "Explaining Religion (Away?): Theism and the Cognitive Science of Religion," *Sophia* 52 (2012): 521–33; Erik Wielenberg, "Evolutionary Debunking Arguments in Religion and Morality," in *Explanation in Ethics and Mathematics*, ed. Uri Leibowitz and

Neil Sinclair (Oxford: Oxford University Press, 2016), 83–102; and Matthew Braddock, "An Evidential Argument for Theism from the Cognitive Science of Religion," in *New Developments in the Cognitive Science of Religion: The Rationality of Religious Belief*, ed. Hans van Eyghen, Rik Peels, and Gijsbert van den Brink (New York: Springer, 2018).

11. Daniel C. Dennett, *Breaking the Spell: Religion as a Natural Phenomenon* (New York: Viking, 2006), 120.

12. Richard Dawkins, *The God Delusion* (Boston: Houghton Mifflin, 2006), 184. Atheistic objectors such as Dawkins and Dennett maintain both that theistic arguments are unsound and that the evidence against God overwhelms what little positive evidence one might find in favor of God. They are evidentialists who claim that the evidence contravenes rational religious belief. Of course, religious evidentialists demur. For the strong evidentialist, theistic or atheistic, the case for rationality/irrationality of belief in God stops here—with an analysis of theistic arguments and counterarguments. In what follows, one might think of the atheist asking this question: Given the prevalence of so much rationally unsupported belief in God, why do so many people believe (irrationally) in God?

13. Such traits, first identified by Stephen Jay Gould and Richard C. Lewontin, are often called spandrels. See their "The Spandrels of San Marco and the Panglossian Paradigm: A Critique of the Adaptationist Programme," *Proceedings of the Royal Society B: Biological Sciences* 205 (1979): 581–98.

14. Noam Chomsky, *The Essential Chomsky* (New York: New Press, 2008), 248.

15. Jesse J. Prinz, "Against Moral Nativism," in *Stich and His Critics*, ed. Michael Bishop and Dominic Murphy (Malden, MA: Wiley-Blackwell, 2009), 168.

16. Prinz, "Against Moral Nativism," 168.

17. I claim only that rational perception or perception-like beliefs require the right sort of causal contact with the object of that belief. I don't claim this of necessary truths.

CHAPTER 5

1. Stephen Stich, *The Fragmentation of Reason* (Cambridge, MA: MIT Press, 1990), 60.

2. Richard Dawkins, *The Blind Watchmaker: Why the Evidence of Evolution Reveals a Universe without Design* (New York: W. W. Norton, 1986), 5.

3. Charles Darwin, letter to William Graham, July 3, 1881, available at https://www.darwinproject.ac.uk/letter/DCP-LETT-13230.xml.

4. Willard Van Orman Quine, "Natural Kinds," in *Ontological Relativity and Other Essays* (New York: Columbia University Press, 1969), 126.

5. Antonia Abbey, "Sex Differences in Attributions for Friendly Behavior: Do Males Misperceive Females' Friendliness?," *Journal of Personality and Social Psychology* 42 (1982): 830–38; Martie G. Haselton, "The Sexual Overperception Bias: Evidence of a Systematic Bias in Men from a Survey of Naturally Occurring Events," *Journal of Research in Personality* 37 (2003): 43–47; Martie G. Haselton and David Buss, "Error Management Theory: A New Perspective on Biases in Cross-Sex Mind Reading," *Journal of Personality and Social Psychology* 78 (2000): 81–91.

6. Stich, *The Fragmentation of Reason*, 62.

7. Friedrich Nietzsche, *The Gay Science: With a Prelude in Rhymes and an Appendix of Songs* (New York: Vintage, 2010), 169.

8. Nietzsche, *The Gay Science*, 169.

9. Patricia Churchland, "Epistemology in the Age of Neuroscience," *Journal of Philosophy* 84.10 (1987): 548–49.

10. Martie G. Haselton and Daniel Nettle, "The Paranoid Optimist: An Integrative Evolutionary Model of Cognitive Biases," *Personality and Social Psychology Review* 10 (2006): 63.

11. Michael Ghiselin, *The Economy of Nature and the Evolution of Sex* (Berkeley: University of California Press, 1974), 126.

12. Haselton and Nettle, "The Paranoid Optimist," 63.

13. Arash Sahraie et al., "Consciousness of the First Order in Blindsight," *Proceedings of the National Academy of Sciences of the United States of America* 49.107 (December 2010): 21217–22; DOI:10.1073/pnas.1015652107.

14. B. Libet et al., "Time of Conscious Intention to Act in Relation to Onset of Cerebral Activity (Readiness-Potential): The Unconscious Initiation of a Freely Voluntary Act," *Brain* 106 (1983): 623–42.

15. While we have raised some questions about Libet-style experiments in the previous chapter, there is no reason to reject the claim that in some or even many cases we are moved to act first and then to form a conscious belief later.

16. Hugo Mercier and Dan Sperber, "Why Do Humans Reason? Arguments for an Argumentative Theory," *Behavioral and Brain Sciences* 34.2 (2011): 57–74.

17. Charles R. Darwin, *The Descent of Man* (London: J. Murray, 1874), 619.

CHAPTER 6

1. Miron Zuckerman, Jordan Silberman, and Judith A. Hall, "The Relation between Intelligence and Religiosity: A Meta-Analysis and Some Proposed Explanations," http://psr.sagepub.com/content/early/2013/08/02/1088868313497266.full.

2. https://www.scientificamerican.com/article/losing-your-religion-analytic-thinking-can-undermine-belief/.

3. http://guardianlv.com/2013/08/atheists-more-intelligent-than-religious-believers-says-new-study/.

4. http://www.medicaldaily.com/proved-atheists-more-intelligent-religious-people-250727.

5. Scott Atran, *In Gods We Trust: The Evolutionary Landscape of Religion* (Oxford: Oxford University Press, 2002), 4.

6. Pascal Boyer, *Religion Explained: The Evolutionary Origins of Religious Thought* (New York: Basic Books, 2001), 2, 330.

7. These studies take analytic thinking as a synonym for inferential thinking (not as philosophers typically take it, as relying on intuitive judgments). Since the intended audience of this essay is philosophers, I will not follow the psychologists and will instead use the philosopher's term "inferential." I will remind the reader throughout of how I am using the terms.

8. Richard Lynn, John Harvey, and Helmuth Nyborg, "Average Intelligence Predicts Atheism Rates across 137 Nations," *Intelligence* 37 (2009): 11–15; all subsequent Lynn quotations are from this short piece.

9. Frederick Hale, "Religious Disbelief and Intelligence: The Failure of a Contemporary Attempt to Correlate National Mean IQs and Rates of Atheism," *Journal for the Study of Religion* 24.1 (2011): 37–53.

10. The idea that IQ somehow measures all that it means to be smart is controversial. Multiple intelligences, including social and emotional intelligence, are missed by traditional IQ.

11. China is looking like a non-Western exception to this trend: as it becomes wealthier, people are becoming more likely to believe in God.

12. Phil Zuckerman, "Atheism: Contemporary Numbers and Patterns," in *The Cambridge Companion to Atheism*, ed. Michael Martin (Cambridge: Cambridge University Press, 2006), 55. See Pippa Norris and Ronald Inglehart, *Sacred and Secular: Religion and Politics Worldwide* (Cambridge: Cambridge University Press, 2004).

13. These correlations are about groups of people, not individual people. Some atheists are surely moved to unbelief by rational reflection.

14. A recent study found that philosophers, though highly trained in critical analysis and moral reflection, were subject to the same biases as non-philosophers when evaluating moral dilemmas. http://www.faculty.ucr.edu/~eschwitz/SchwitzPapers/Stability-150 423.pdf.

15. Solomon E. Asch, "Opinions and Social Pressure," *Scientific American* 193.5 (1955): 31–35; and Asch, "Studies of Independence and Conformity, I: A Minority of One against a Unanimous Majority," *Psychological Monographs: General and Applied* 70.9 (1956): 1–70.

16. I endorse methodological naturalism. See Kelly James Clark, "Atheism and Analytic Thinking," in *The Science and Religion Dialogue: Past and Future*, ed. Michael Welker (New York: Peter Lang, 2014), 245–56. Methodological naturalism maintains that, in the practice of science, scientists should not appeal to supernatural entities (like gods or ghosts) or forces (like qi). However, "methodological" neither assumes nor implies ontological naturalism (the claim that supernatural entities, like God and souls, don't exist).

17. In addition, these biases engender and perpetuate hiring biases.

18. Gordon Hodson and Michael A. Busseri, "Bright Minds and Dark Attitudes: Lower Cognitive Ability Predicts Greater Prejudice through Right-Wing Ideology and Low Intergroup Contact," *Psychological Science* 23 (2012): 187–95. One reason to be cautious about how to interpret IQ test scores is that familiarity with the type of test and certain acquired ways of thinking (due to social environment, education, etc.) can lead to measurement error.

19. Samuel Goldman, "Why Isn't My Professor Conservative?," *American Conservative*, January 7, 2016, http://www.theamericanconservative.com/articles/why-isnt-my-professor -conservative/.

20. Bruce Charlton attributes left-wing political views among academics to "over-use of abstraction." Charlton, "Clever Sillies: Why High IQ People Tend to Be Deficient in Common Sense," *Medical Hypotheses* 73.6 (2009): 867–70.

21. We are speaking in terms of general tendencies, not cognitive necessities. So, for example, while we (the entire group of human beings) may be generally inclined toward intuitive religious belief, not everyone will be a religious believer, and not every religious believer will have acquired his or her beliefs noninferentially. The claim that we (as humans) are typically natural and nonreflective theists is consistent with there being atheists and inferential theists.

22. http://abcnews.go.com/blogs/health/2012/04/26/logic-linked-to-religious-disbelief -study-implies/.

23. http://news.sciencemag.org/2012/04/keep-faith-dont-get-analytical/.

24. http://www.scientificamerican.com/article/how-critical-thinkers-lose-faith-god/.

25. http://theconversation.com/analytic-thinking-erodes-religious-belief-6709/.

26. http://www.pewresearch.org/fact-tank/2013/10/23/5-facts-about-atheists/.

27. http://redcresearch.ie/wp-content/uploads/2012/08/RED-C-press-release-Religion -and-Atheism-25-7-12.pdf.

28. Robert N. McCauley, *Why Religion Is Natural and Science Is Not* (Oxford: Oxford University Press, 2011).

29. Given our repeated relapses into folk physics, one might think that we can never fully overcome our natural dispositions.

30. Again, I am speaking in generalities. One might believe $E = mc^2$ because one was told it, not as a result of inferential thinking (though I doubt, under such circumstances, one would understand it well at all). Moreover, one might be an atheist because one's parents taught one at the earliest age that there was no God (which required no inferential thinking on one's part).

31. Will M. Gervais and Ara Norenzayan, "Analytic Thinking Promotes Religious Disbelief," *Science* 336 (2012): 493–96.

32. Although Gervais and Norenzayan's studies prompted the headlines, they themselves resisted the sensational conclusions of the preceding section. They write: "Finally, we caution that the present studies are silent on long-standing debates about the intrinsic value or rationality of religious beliefs, or about the relative merits of analytic and intuitive thinking in promoting optimal decision making" (Gervais and Norenzayan, "Analytic Thinking Promotes Religious Disbelief," 496).

33. Shane Frederick, "Cognitive Reflection and Decision Making," *Journal of Economic Perspectives* 19 (2005): 25–42.

34. The quick and easy intuitive yet wrong response to (1) is 10 cents, while the correct analytic, deliberate answer is 5 cents; to (2) is 100 while the analytic answer is 5; and to (3) is 24 while the analytic is 47.

35. Gervais and Norenzayan, "Analytic Thinking Promotes Religious Disbelief," 494.

36. The other studies involved implicit primes and art primes. Implicit primes involved arranging words into sentences; the prime group was given thinking terms ("reason," "analyze," "ponder," etc.), while the control group was given unrelated words ("hammer," "shoe," "jump," etc.). Participants in the art control group stared at a "neutral" image such as *The Discobolos*, whereas the remainder was primed by staring at *The Thinker* (an "artwork depicting a reflective thinking pose").

37. Amitai Shenhav, David G. Rand, and Joshua D. Greene, "Divine Intuition: Cognitive Style Influences Belief in God," *Journal of Experimental Psychology: General* 141 (2012): 423–28.

38. Gordon Pennycock et al., "Analytic Cognitive Style Predicts Religious and Paranormal Belief," *Cognition* 123.3 (2012): 335–46.

39. Pennycock et al., "Analytic Cognitive Style Predicts Religious and Paranormal Belief," 339.

40. Gervais and Norenzayan, "Analytic Thinking Promotes Religious Disbelief."

41. Joseph Henrich, "The Evolution of Costly Displays, Cooperation, and Religion: Credibility Enhancing Displays and Their Implications for Cultural Evolution," *Evolution and Human Behavior* 30 (2009): 244–60.

42. Joseph Bulbulia, "Religious Costs as Adaptations That Signal Altruistic Intention," *Evolution and Cognition* 10 (2004): 19–38; and Richard Sosis, "Does Religion Promote Trust? The Role of Signaling, Reputation, and Punishment," *Interdisciplinary Journal of Research on Religion* 1 (2005): 1–30.

43. Gervais and Norenzayan, "Analytic Thinking Promotes Religious Disbelief," 495.

44. Here is a tendentious way of putting it: suppose that, instead of a correlation between atheism and such cognitive goods, there is instead a connection between atheism and a cognitive defect.

45. Moral beliefs can likewise be affected by various cognitive attitudes and deficiencies. The cool calculations required of utilitarianism have been associated with "egocentric attitudes and less identification with humanity" and are negatively correlated with altruism. See Guy Kahane, "Evolutionary Debunking Arguments," *Noûs* 45 (2011): 103–25. Other studies suggest that utilitarian judgment is associated with antisocial traits such as psychopathy. See Daniel M. Bartels and David A. Pizarro, "The Mismeasure of Morals: Antisocial Personality Traits Predict Utilitarian Responses to Moral Dilemmas," *Cognition* 121.1 (2011): 154–61; Andrea L. Glenn et al., "Moral Identity in Psychopathy," *Judgment and Decision Making* 5 (2010): 497–505; Michael Koenigs et al., "Utilitarian Moral Judgment in Psychopathy," *Social, Cognitive and Affective Neuroscience* 7.6 (2012): 708–14; and Katje Wiech et al., "Cold or Calculating? Reduced Activity in the Subgenual Cingulate Reflects Decreased Aversion to Harming in Counterintuitive Utilitarian Judgment," *Cognition* 126 (2013): 364–72. Studies suggest that utilitarian judgment is also associated with antisocial traits such as diminished empathic concern. See So Young Choe and Kyung-Hwan Min, "Who Makes Utilitarian Judgments? The Influences of Emotions on Utilitarian Judgments," *Judgment and Decision Making* 6 (2011): 580–92; and Molly J. Crockett et al., "Serotonin Selectively Influences Moral Judgment and Behavior through Effects on Harm Aversion," *Proceedings of the National Academy of Sciences of the United States of America* 107 (2010): 17433–38.

46. High-functioning autism (HFA) is a diagnosis that overlaps with Asperger's syndrome. Compared to the earlier (DSM IV) diagnosis of autistic disorder, individuals in both groups have average or above-average intelligence but may differ in age of onset, early language development, and motor skills. Catherine L. Caldwell-Harris, Caitlin Fox Murphy, and Tessa Velazquez, "Religious Belief Systems of Persons with High Functioning Autism," in *Proceedings of the 33rd Annual Meeting of the Cognitive Science Society* (Austin: Cognitive Science Society, 2011), 3362–66.

47. Norenzayan's Autism Studies 3 and 4 were conducted with a broad sample of Americans. Ara Norenzayan, Will M. Gervais, and Kali H. Trzesniewski, "Mentalizing Deficits Constrain Belief in a Personal God," *PLOS One* 7.5 (2012). In their Study 1, they found that "autistic participants were only 11% as likely as neurotypical controls to strongly endorse God."

48. We should be warned at the outset of rushing to judgment here. Since autism is so complex, it is probably impossible to say that autism per se mediates atheism. At best, studies might isolate a specific quality that is common in autism, such as mentalizing constraints. Moreover, as seen in the previous sections, unbelief in a personal God is not tantamount to atheism. Atheism would require unbelief in both personal and impersonal-but-transcendent (and perhaps moral) forces.

49. The three diagnostic tests were Baron-Cohen's autism quotient, the "reading the mind in the eyes" test, and the systemizing quotient. For the first, see Simon Baron-Cohen et al., "The Autism-Spectrum Quotient (AQ): Evidence from Asperger Syndrome/High-Functioning Autism, Males and Females, Scientists and Mathematicians," *Journal of Autism and Developmental Disorders* 31.1 (2001): 5–17. For the others, see Simon Baron-Cohen et al., "The Systemizing Quotient: An Investigation of Adults with Asperger's Syndrome or High Functioning Autism, and Normal Sex Differences," *Philosophical Transactions of the Royal Society* 358 (2003): 361–74.

50. Let me offer some cautionary notes at this point. Questionnaires don't reveal nuances and ambiguities. For example, these studies often fail to distinguish between personal and

impersonal gods. Since atheism is unbelief in both personal and impersonal gods, offering or suggesting that atheism is simply unbelief in a personal god is too blunt an instrument to access more fine-grained beliefs (and unbeliefs). Moreover, while some people may say, "I am not religious" in questionnaires, in personal interviews they report religious aspects of their lives. One rarely learns of such subtleties through questionnaires.

51. Recent studies aren't so confident of the atheist-autism connection. See Rachel S. Brezis, "Autism as a Case for Neuroanthropology: Delineating the Role of Theory of Mind in Religious Development," in *The Encultured Brain: An Introduction to Neuroanthropology*, ed. Daniel H. Lende and Greg Downey (Cambridge, MA: MIT Press, 2012), 291–31; Hanneke Schaap-Jonker et al., "Autism Spectrum Disorders and the Image of God as a Core Aspect of Religiousness," *International Journal for the Psychology of Religion* 23.2 (2012): 145–60; Paul Reddish, Penny Tok, and Radek Kundt, "Religious Cognition and Behaviour in Autism: The Role of Mentalizing," *International Journal for the Psychology of Religion* (2015): 95–112.

52. To be sure, autism spectrum disorder (ASD) comes in degrees. It can range from the profound social, language, and behavioral problems that are characteristic of autistic disorder to the milder Asperger's syndrome. These correlations aren't claiming, for example, that all or most scientists are fully autistic. The claims in this paragraph are based on Baron-Cohen et al., "The Autism-Spectrum Quotient (AQ)"; Baron-Cohen, Richler, Bisarya, Gurunathan, and Wheelwright, "The Systemizing Quotient"; and Simon Baron-Cohen et al., "Mathematical Talent Is Linked to Autism," *Human Nature* 18 (2007): 125–31.

53. Elaine Eklund has the best work on religious belief among scientists. For a summary, see http://religion.ssrc.org/reforum/Ecklund.pdf.

CHAPTER 7

1. Gijsbert Stoet and David C. Geary, "Sex Differences in Mathematics and Reading Achievement Are Inversely Related: Within- and Across-Nation Assessment of 10 Years of PISA Data," *PLOS One* (2013).

2. Yu Xie and Kimberlee A. Shauman, *Women in Science: Career Processes and Outcomes* (Cambridge, MA: Harvard University Press, 2003).

3. According to some studies, people have been shown to "play down" to stereotypes. For example, if women are aware of the stereotype concerning math, they perform more poorly on math tests. Moreover, white males will perform more poorly if they think they are being compared to Asian males than if they think they are being compared to other white guys. So performance can get trapped in a vicious cycle: a slight weakness in an area (e.g., due to differential treatment at home that has been historically inherited) can create a small but noticeable difference that spawns a stereotype that then leads to playing down to the stereotype that accentuates the difference, and so on.

4. Ara Norenzayan, Will M. Gervais, and Kali H. Trzesniewski, "Mentalizing Deficits Constrain Belief in a Personal God," *PLOS One* 7.5 (2012): e36880.

5. I am grateful to Ingela Visuri for her helpful comments on this chapter.

6. Mark Haddon, *The Curious Incident of the Dog in the Night-Time* (New York: Doubleday, 2003), 2–3.

7. American Psychiatric Association, *The Diagnostic and Statistic Manual of Mental Disorders*, 5th ed. (Arlington: American Psychiatric Association, 2013), referred to as the DSM 5.

The autism spectrum was earlier formulated as four separate diagnoses in the DSM IV: *autistic disorder*, *Asperger's disorder* (often referred to as Asperger's syndrome, or AS), *childhood disintegrative disorder*, and *pervasive developmental disorder otherwise not specified*. These were replaced by *autism spectrum disorder* (ASD), an umbrella term that is supposed to include all the earlier diagnoses, from mild to severe symptoms. Naming autism a *disorder* has been understood as pejorative, especially within the autism community, and the term is occasionally replaced by *autism spectrum condition* (ASC), also in this text. There is also a move toward talking of *autisms* in the plural, acknowledging the fact that individuals develop very differently.

8. To avoid the misconception that individuals on the spectrum lack empathy, it is important to note that affective empathy seems intact, or possibly even heightened, in some autistic individuals. See Adam Smith, "The Empathy Imbalance Hypothesis of Autism: A Theoretical Approach to Cognitive and Emotional Empathy in Autistic Development," *Psychological Record* 59.3 (2009): 489–510. In their article "Mixed Emotions: The Contribution of Alexithymia to Emotional Symptoms of Autism," *Translational Psychiatry* 3 (2013): 1–8, Geoff Bird and Richard Cook suggest that emotional "blindness" in some individuals may in fact be due to alexithymia, rather than being a primary feature of autism. Alexithymia is a personality characteristic in which the individual is unable to identify and describe emotions in the self. The main features of alexithymia are dysfunction in emotional awareness, social attachment, and interpersonal relating.

9. http://socialintelligence.labinthewild.org/mite/.

10. Simon Baron-Cohen, Alan M. Leslie, and Uta Frith, "Does the Autistic Child Have a 'Theory of Mind'?," *Cognition* 21 (1985): 37–46.

11. Baron-Cohen has developed the systemizing-quotient (SQ) to assess this.

12. Nick Dubin and Janet E. Graetz, "Through a Different Lens: Spirituality in the Lives of Individuals with Asperger's Syndrome," *Journal of Religion, Disability and Health* 13.1 (2009): 29–39. See also Ingela Visuri, "Could Everyone Talk to God? A Case Study on Asperger's Syndrome, Religion and Spirituality," *Journal of Religion, Health and Disability* 16.4 (2012): 352–78.

13. Simon Baron-Cohen and Sally Wheelwright, "Obsessions in Children with Autism or Asperger Syndrome: A Content Analysis in Terms of Core Domains of Cognition," *British Journal of Psychiatry* 175 (1999): 484–90.

14. The prevalence of savant abilities among autistic individuals, the stuff of Rain Man, is about 10 percent, considerably higher than the 1 percent of the entire population.

15. I will use "mentalizing" rather than "ToM" throughout this chapter. Some autism researchers reject ToM because they believe that it assumes the mirror neuron hypothesis (although I make no such assumption in my understanding of ToM). There is also a trend toward saying that autistic people cannot mentalize because of sensory overload. "Mentalizing" is the broader term for an everyday ability, while "ToM" is related to specific cognitive abilities.

16. Justin Barrett, *Why Would Anyone Believe in God?* (Lanham: AltaMira, 2004).

17. http://www.hbo.com/movies/temple-grandin.

18. Grandin's religiosity is neither representative nor archetypal of an autistic individual's spirituality. It is also not untypical.

19. Temple Grandin, *Thinking in Pictures: My Life with Autism*, expanded ed. (New York: Vintage Books, 2006), 224.

20. Grandin, *Thinking in Pictures*, 231.

21. Jesse M. Bering, *The God Instinct: The Psychology of Souls, Destiny, and the Meaning of Life* (London: Nicholas Brealey, 2011), 85.

22. Grandin, *Thinking in Pictures*, 233.

23. This requires some ability to mentalize. Perhaps since social signals from animals are less contradictory, less "noisy," than those from human beings, she is able to mentalize successfully with them. See Smith, "The Empathy Imbalance Hypothesis of Autism," for an interesting discussion on cognitive and emotional empathy in autistic individuals.

24. Anne Raver, "Qualities of an Animal Scientist: Cow's Eye View and Autism," *New York Times*, August 5, 1997, https://www.nytimes.com/1997/08/05/science/qualities-of-an-animal-scientist-cow-s-eye-view-and-autism.html.

25. This is one form of synesthesia, a mixing of the senses, in which objects such as letters, shapes, or numbers are conceived of as a sensory perception such as smell, color, or flavor.

26. Daniel Tammet, *Born on a Blue Day: Inside the Extraordinary Mind of an Autistic Savant* (New York: Free Press, 2007), 2.

27. Tammet, *Born on a Blue Day*, 4–5.

28. Tammet, *Born on a Blue Day*, 223.

29. Tammet, *Born on a Blue Day*, 225.

30. Tammet, *Born on a Blue Day*, 225.

31. Tony Attwood, *The Complete Guide to Asperger's Syndrome* (London: Jessica Kingsley, 2007).

32. Smith, "The Empathy Imbalance Hypothesis of Autism."

33. Tammet, *Born on a Blue Day*, 67.

34. Tammet, *Born on a Blue Day*, 223.

35. Dubin and Graetz, "Through a Different Lens"; Robert N. McCauley, *Why Religion Is Natural and Science Is Not* (Oxford: Oxford University Press, 2011).

36. Scott Bellini, "Social Skill Deficits and Anxiety in High-Functioning Adolescents with Autism Spectrum Disorders," *Focus on Autism and Other Developmental Disabilities* 19 (2004): 78–86; and Scott Bellini, "The Development of Social Anxiety in Adolescents with Autism Spectrum Disorders," *Focus on Autism and Other Developmental Disabilities* 21 (2006): 138–45.

37. Tammet, *Born on a Blue Day*, 75.

38. Keise Izuma, Kenji Matsumoto, Colin F. Camerer, and Ralph Adolphs, "Insensitivity to Social Reputation in Autism," *Proceedings of the National Academy of Sciences of the United States of America* 108.42 (2011): 17302–7.

39. Dubin and Graetz, "Through a Different Lens."

CHAPTER 8

1. Elizabeth Spelke and Katherine D. Kinzler, "Core Knowledge," *Developmental Science* 11 (2007): 89–96. This sort of research relies on subtle behavioral cues such as eye gaze to determine what infants "know" or expect. For instance, if, given two different displays, babies preferentially attend to one versus the other, scientists infer that babies can discriminate between the two displays. Similarly, if babies watch one display until their attention is lost (they stop looking at it), and a second display is then presented that recovers the babies' attention, scientists infer that babies notice a difference in the second display. In research on infants' understandings of physical objects, babies might be shown a display in which a ball rolls down a ramp from the right to the left, disappearing behind an opaque screen and then reappearing on the other side. The display is repeated over and over until the baby becomes "habituated"

(i.e., bored). Then babies might be shown the same display with the screen removed. Generally, such a display does not recapture infants' attention, apparently because it only depicts what they assumed was going on previously: it presents nothing new. But when the display is changed for comparison, when a barrier previously hidden by the screen is revealed—a barrier that, from an adult perspective, would clearly block the motion of the rolling ball—babies' attention is more likely to be recovered. Scientists infer that babies "know" that balls cannot roll through solid barriers. Babies find the new information about the presence of a solid obstacle surprising. Research of this kind, then, gives evidence that preverbal babies hold a host of expectations about objects in their environments: when babies recognize something as a bounded, physical object (as opposed to a pile of sand or a cloud), they automatically, noninferentially expect a range of properties to apply to the object.

2. Lawrence A. Hirschfeld and Susan A. Gelman, eds., *Mapping the Mind: Domain Specificity in Cognition and Culture* (Cambridge: Cambridge University Press, 1994), and Dan Sperber, Francesco Cara, and Vittorio Girotto, "Relevance Theory Explains the Selection Task," *Cognition* 57 (1995): 31–95, include examples of scientific research in these areas. Pascal Boyer and Pierre Lienard, "Whence Collective Rituals? A Cultural Selection Model of Ritualized Behavior," *American Anthropologist* 108 (2006): 814–28, summarize evidence for a hazard-precaution faculty and discuss its potential ability to explain some dynamics of cultural rituals.

3. For a full discussion of these issues, see Jonathan Haidt, *The Righteous Mind: Why Good People Are Divided by Politics and Religion* (New York: Pantheon Books, 2012).

4. Daniel Dennett, *Darwin's Dangerous Idea: Evolution and the Meanings of Life* (New York: Simon and Schuster, 1995).

5. There are complex socioeconomic factors involved as well. Rich people, or people driving fancier cars, are more likely to believe they are privileged with respect to the law. See Elizabeth Norton, "Shame on the Rich," *Science*, February 27, 2012, http://news.sciencemag .org/2012/02/shame-rich.

APPENDIX

1. Richard Dawkins, *A Devil's Chaplain: Reflections on Hope, Lies, Science, and Love* (Boston: Houghton Mifflin, 2003), 117.

2. Alexandra L. Varga and Kai Hamburger, "Beyond Type 1 vs. Type 2 Processing: The Tri-Dimensional Way," *Frontiers in Psychology* 5 (2014): 993.

3. Kelly James Clark, *Return to Reason* (Grand Rapids: Eerdmans, 1990); Elizabeth Spelke and Katherine D. Kinzler, "Core Knowledge," *Developmental Science* 11 (2007): 89–96; Kelly James Clark and Justin L. Barrett, "Reidian Religious Epistemology and the Cognitive Science of Religion," *Journal of the American Academy of Religion* 79.3 (2011): 639–75.

4. I am not insensitive to intuitive biases, which have been well documented; see Daniel Kahneman, *Thinking Fast and Slow* (New York: Farrar, Straus and Giroux, 2011). But there are also inferential biases. For example, we tend to be sensitive to evidence or arguments that support our beliefs and to be insensitive to evidence or arguments that are contrary to our beliefs. Not all inferential beliefs are true. People have inferred such untrue beliefs as the phlogiston theory, "women should aspire to be beautiful" (since they cannot be rational), and "Nixon will make a great president." Scientists seem to have inferred themselves into a contradiction between science's two most widely accepted and successful theories—quantum

mechanics and general relativity. They cannot both be true. Finally, philosophers, among the most ardent defenders of argument, continue to hold a wide diversity of incompatible beliefs. Some philosophers believe enthusiastically while others deny with equal vehemence the following and more (I take just a few claims in ethics; examples could be drawn from every area of philosophy): there are moral absolutes, there are moral facts, and there is human virtue.

5. Some philosophers contend that philosophical intuitions have evidential value, while others ardently reject that contention. See Herman Cappelen, *Philosophy without Intuitions* (Oxford: Oxford University Press, 2012). There is increasing empirical evidence that intuitions vary according to, for example, cultural background, socioeconomic status, and affective state. See Jonathan M. Weinberg, Shaun Nichols, and Stephen Stich, "Normativity and Epistemic Intuitions," *Philosophical Topics* 29 (2001): 429–60; Shaun Nichols, Stephen Stich, and Jonathan M. Weinberg, "Metaskepticism: Meditations in Ethno-Epistemology," in *The Skeptics*, ed. Stephen Luper (Aldershot: Ashgate, 2003), 227–47; Edouard Machery et al., "Semantics, Cross-Cultural Style," *Cognition* 92.3 (2004): B1–B12; Shaun Nichols and Joshua Knobe, "Moral Responsibility and Determinism: The Cognitive Science of Folk Intuitions," *Noûs* 41.4 (2007): 663–85; Stacey Swain, Joshua Alexander, and Jonathan M. Weinberg, "The Instability of Philosophical Intuitions: Running Hot and Cold on True Temp," *Philosophy and Phenomenological Research* 76.1 (2008): 138–55.

6. We must also assume (take as given) various epistemic principles about the nature and normativity of belief. For example, one must assume (or reject) (a) that belief in God must be based on evidence, or (b) that disagreement among those who are one's intellectual equals undermines one's rationality. If one affirms (a), one must also make assumptions about the nature of argument—deductive, probabilistic, cumulative case, inference to best explanation.

7. I leave it for the strong evidentialist, atheist or theist, to defend the veracity of their basic intuitions involved in theistic arguments; as one might guess from my discussion, I am dubious.

8. Hilary Kornblith, *On Reflection* (Oxford: Oxford University Press, 2012).

9. Jerry Kang and Kristin Lane, "Seeing Through Colorblindness: Implicit Bias and the Law," *UCLA Law Review* 58.2 (2010): 467.

10. Eldar Shafir, ed., *The Behavioral Foundations of Public Policy* (Princeton: Princeton University Press, 2012).

11. Kornblith, *On Reflection*, 24–25.

12. It is hard to imagine a plausible (naturalistic) evolutionary story in which developing reliable philosophical intuitions was reproductively successful.

13. Lest one think I equally disparage all inference, I don't. While I believe that philosophical and theological arguments resolve into, at bottom, unshared intuitions about the nature of reality, I don't mean that every argument in every field does. For example, I think we have widely shared intuitions that guide thinking in both mathematics and the natural sciences. I don't claim that science is intuition-free. It is not. But since scientists today share many of those intuitions, scientific debates aren't irresolvable.

Bibliography

Abbey, Antonia. "Sex Differences in Attributions for Friendly Behavior: Do Males Misperceive Females' Friendliness?" *Journal of Personality and Social Psychology* 42 (1982): 830–38.

Alexander, Eben. *Proof of Heaven: A Neurosurgeon's Journey into the Afterlife.* New York: Simon and Schuster, 2012.

Alexander, Joshua, and Jonathan Weinberg. "Analytic Epistemology and Experimental Philosophy." *Philosophy Compass* 2 (2007): 56–80.

Alper, Matthew. *The God Part of the Brain.* Boulder: Rogue, 2000.

Alston, William. *Perceiving God: The Epistemology of Religious Experience.* Ithaca: Cornell University Press, 1991.

———. *The Reliability of Sense Perception.* Ithaca: Cornell University Press, 1993.

American Psychiatric Association. *Diagnostic and Statistic Manual of Mental Disorders.* 5th ed. Arlington: American Psychiatric Publishing, 2013.

Apperly, Ian. *Mindreaders: The Cognitive Basis of Theory of Mind.* Hove: Psychology Press, 2011.

Apperly, Ian, and Stephen A. Butterfill. "Do Humans Have Two Systems to Track Beliefs and Belief-like States?" *Psychological Review* 116.4 (2009): 953–70.

Argyle, Michael. *Psychology and Religion.* London: Routledge, 2000.

Aron, Arthur, Elizabeth Nancy Aron, and Danny Smollan. "Inclusion of Other in the Self Scale and the Structure of Interpersonal Closeness." *Journal of Personality and Social Psychology* 63.4 (1992): 596–612.

Asch, Solomon E. "Opinions and Social Pressure." *Scientific American* 193.5 (1955): 31–35.

———. "Studies of Independence and Conformity, I: A Minority of One against a Unanimous Majority." *Psychological Monographs: General and Applied* 70.9 (1956): 1–70.

Atran, Scott. *In Gods We Trust: The Evolutionary Landscape of Religion.* Oxford: Oxford University Press, 2002.

Attwood, Tony. *The Complete Guide to Asperger's Syndrome*. London: Jessica Kingsley, 2007.

Audi, Robert. *Epistemology: A Contemporary Introduction to the Theory of Knowledge*. New York: Routledge, 2010.

Baron-Cohen, Simon. "Autism: The Empathizing-Systemizing (E-S) Theory." *Annals of the New York Academy of Sciences* 1156.1 (2009): 68–80.

———. "The Autistic Child's Theory of Mind—A Case of Specific Developmental Delay." *Journal of Child Psychology and Psychiatry and Allied Disciplines* 30.2 (1989): 285–97.

———. *Mindblindness: An Essay on Autism and Theory of Mind*. Cambridge, MA: MIT Press, 1995.

Baron-Cohen, Simon (principal author). *Mind Reading: The Interactive Guide to Emotions*. London: Jessica Kingsley Publisher, 2003.

Baron-Cohen, Simon, Rosa A. Hoekstra, Rebecca Knickmeyer, and Sally Wheelwright. "The Autism-Spectrum Quotient (AQ)—Adolescent Version." *Journal of Autism and Developmental Disorders* 36.3 (2006): 343–50.

Baron-Cohen, Simon, Therese Jolliffe, Catherine Mortimore, and Mary Robertson. "Another Advanced Test of Theory of Mind: Evidence from Very High Functioning Adults with Autism or Asperger Syndrome." *Journal of Child Psychology and Psychiatry and Allied Disciplines* 38.7 (1997): 813–22.

Baron-Cohen, Simon, Alan M. Leslie, and Uta Frith. "Does the Autistic Child Have a 'Theory of Mind'?" *Cognition* 21 (1985): 37–46.

Baron-Cohen, Simon, Jennifer Richler, Dheraj Bisarya, Nhishanth Gurunathan, and Sally Wheelwright. "The Systemizing Quotient: An Investigation of Adults with Asperger's Syndrome or High Functioning Autism, and Normal Sex Differences." *Philosophical Transactions of the Royal Society* 358 (2003): 361–74.

Baron-Cohen, Simon, Fiona J. Scott, Carrie Allisona, Joanna G. Williams, Patrick Bolton, Fiona Elaine Matthews, and Carol Brayne. "Prevalence of Autism-Spectrum Conditions: UK School-Based Population Study." *British Journal of Psychiatry* 194.6 (2009): 500–509.

Baron-Cohen, Simon, and Sally Wheelwright. "The Empathy Quotient: An Investigation of Adults with Asperger Syndrome or High Functioning Autism, and Normal Sex Differences." *Journal of Autism and Developmental Disorders* 34.2 (2004): 163–75.

Baron-Cohen, Simon, and Sally Wheelwright. "Obsessions in Children with Autism or Asperger Syndrome: A Content Analysis in Terms of Core Domains of Cognition." *British Journal of Psychiatry* 175 (1999): 484–90.

Baron-Cohen, Simon, Sally Wheelwright, Amy Burtenshaw, and Esther Hobson. "Mathematical Talent Is Linked to Autism." *Human Nature* 18 (2007): 125–31.

Baron-Cohen, Simon, Sally Wheelwright, Richard Skinner, Joanne Martin, and Emma Clubley. "The Autism-Spectrum Quotient (AQ): Evidence from Asperger Syndrome/High-Functioning Autism, Males and Females, Scientists and Mathematicians." *Journal of Autism and Developmental Disorders* 31.1 (2001): 5–17.

Barrett, Justin L. *Born Believers: The Science of Children's Religious Belief.* New York: Simon and Schuster, 2012.

———. *Cognitive Science, Religion, and Theology: From Human Minds to Divine Minds.* West Conshohocken, PA: Templeton, 2011.

———. "Theological Correctness: Cognitive Constraint and the Study of Religion." *Method and Theory in the Study of Religion* 11 (1999): 325–39.

———. *Why Would Anyone Believe in God?* Lanham: AltaMira, 2004.

Barrett, Justin L., and Frank C. Keil. "Conceptualizing a Nonnatural Entity: Anthropomorphism in God Concepts." *Cognitive Psychology* 31 (1996): 219–47.

Barrett Justin L., and Melanie Nyhof. "Spreading Non-Natural Concepts: The Role of Intuitive Conceptual Structures in Memory and Transmission of Cultural Materials." *Journal of Cognition and Culture* 1 (2001): 69–100.

Barrett, Justin L., and Rebekah A. Richert. "Anthropomorphism or Preparedness? Exploring Children's God Concepts." *Review of Religious Research* 44.3 (2003): 300–312.

Bartels, Daniel M., and David A. Pizarro. "The Mismeasure of Morals: Antisocial Personality Traits Predict Utilitarian Responses to Moral Dilemmas." *Cognition* 121.1 (2011): 154–61.

Bauminger, Nirit, and Connie Kasari. "Brief Report: Theory of Mind in High-Functioning Children with Autism." *Journal of Autism and Developmental Disorders* 29.1 (1999): 81–86.

Bellini, Scott. "The Development of Social Anxiety in Adolescents with Autism Spectrum Disorders." *Focus on Autism and Other Developmental Disabilities* 21 (2006): 138–45.

———. "Social Skill Deficits and Anxiety in High-Functioning Adolescents with Autism Spectrum Disorders." *Focus on Autism and Other Developmental Disabilities* 19 (2004): 78–86.

Bergman, Michael. *Justification without Awareness.* Oxford: Oxford University Press, 2006.

Bering, Jesse M. *The Belief Instinct: The Psychology of Souls, Destiny, and the Meaning of Life.* New York: W. W. Norton, 2011.

———. "The Existential Theory of Mind." *Review of General Psychology* 6 (2002): 3–24.

———. "The Folk Psychology of Souls." *Behavioral and Brain Sciences* 29 (2006): 453–62.

———. *The God Instinct: The Psychology of Souls, Destiny, and the Meaning of Life*. London: Nicholas Brealey, 2011.

———. "Intuitive Conceptions of Dead Agents' Minds: The Natural Foundations of Afterlife Beliefs as Phenomenological Boundary." *Journal of Cognition and Culture* 2.4 (2002): 263–308.

Bering, Jesse M., and Dominic Johnson. "'O Lord . . . You Perceive My Thoughts from Afar': Recursiveness and the Evolution of Supernatural Agency." *Journal of Cognition and Culture* 5 (2005): 118–42.

Bird, Geoff, and Richard Cook. "Mixed Emotions: The Contribution of Alexithymia to Emotional Symptoms of Autism." *Translational Psychiatry* 3 (2013): 1–8.

Bloom, Paul. *Descartes' Baby: How Child Development Explains What Makes Us Human*. London: William Heinemann, 2004.

———. "Is God an Accident?" *Atlantic Monthly*, December 2005.

———. "Religion Is Natural." *Developmental Science* 10 (2007): 147–51.

———. "Religious Belief as an Evolutionary Accident." In *The Believing Primate: Scientific, Philosophical, and Theological Reflections on the Origin of Religion*, edited by Jeffrey Schloss and Michael Murray, 118–27. Oxford: Oxford University Press, 2009.

———. "Religious Thought and Behaviour as By-Products of Brain Function." *Trends in Cognitive Sciences* 7.3 (2003): 119–24.

Borjigin, Jimo, UnCheol Lee, Tiecheng Liu, Dinesh Pal, Sean Huff, Daniel Klarr, Jennifer Sloboda, Jason Hernandez, Michael M. Wang, and George A. Mashour. "Surge of Neurophysiological Coherence and Connectivity in the Dying Brain." *Proceedings of the National Academy of Sciences of the United States of America* 110.35 (2013): 14432–37.

Boucher, Jill. "Memory and Generativity in Very High Functioning Autism: A Firsthand Account and an Interpretation." *Autism* 11 (2007): 255–64.

Bowler, D. M. "Theory of Mind in Aspergers Syndrome." *Journal of Child Psychology and Psychiatry and Allied Disciplines* 33.5 (1992): 877–93.

Boyer, Pascal. *The Naturalness of Religious Ideas: A Cognitive Theory of Religion*. Berkeley: University of California Press, 1994.

———. *Religion Explained: The Evolutionary Origins of Religious Thought*. New York: Basic Books, 2001.

———. "Why Is Religion Natural?" *Skeptical Inquirer* 28.2 (March/April 2004).

Boyer, Pascal, and Pierre Lienard. "Whence Collective Rituals? A Cultural Selection Model of Ritualized Behavior." *American Anthropologist* 108 (2006): 814–28.

Braddock, Matthew. "An Evidential Argument for Theism from the Cognitive Science of Religion." In *New Developments in the Cognitive Science of Religion: The Rationality of Religious Belief*, edited by Hans van Eyghen, Rik Peels, and Gijsbert van den Brink. New York: Springer, 2018.

Brezis, Rachel S. "Autism as a Case for Neuroanthropology: Delineating the Role of Theory of Mind in Religious Development." In *The Encultured Brain: An Introduction to Neuroanthropology*, edited by Daniel H. Lende and Greg Downey, 291–314. Cambridge, MA: MIT Press, 2012.

Bulbulia, Joseph. "The Evolution of Religion." In *Oxford Handbook of Evolutionary Psychology*, edited by Robin Ian MacDonald Dunbar and Louise Barrett, 621–36. Oxford: Oxford University Press, 2007.

———. "Nature's Medicine: Religiosity as an Adaptation for Health and Cooperation." In *Where God and Science Meet: How Brain and Evolutionary Studies Alter Our Understanding of Religion*, 3 vols., edited by Patrick McNamara, 1:81–121. Westport: Praeger, 2006.

———. "Religious Costs as Adaptations That Signal Altruistic Intention." *Evolution and Cognition* 10 (2004): 19–38.

———, ed. *The Evolution of Religion: Studies, Theories, and Critiques*. Santa Margarita, CA: Collins Foundation, 2008.

Caldwell-Harris, Catherine L. "Understanding Atheism/Non-belief as an Expected Individual-Differences Variable." *Religion, Brain and Behavior* 2.1 (2012): 4–47.

Caldwell-Harris, Catherine L., Caitlin Fox Murphy, and Tessa Velazquez. "Religious Belief Systems of Persons with High Functioning Autism." In *Proceedings of the 33rd Annual Meeting of the Cognitive Science Society*, 3362–66. Austin: Cognitive Science Society, 2011.

Cappelen, Herman. *Philosophy without Intuitions*. Oxford: Oxford University Press, 2012.

Charlton, Bruce. "Clever Sillies: Why High IQ People Tend to Be Deficient in Common Sense." *Medical Hypotheses* 73.6 (2009): 867–70.

Choe, So Young, and Kyung-Hwan Min. "Who Makes Utilitarian Judgments? The Influences of Emotions on Utilitarian Judgments." *Judgment and Decision Making* 6 (2011): 580–92.

Chomsky, Noam. *The Essential Chomsky*. New York: New Press, 2008.

———. *Rules and Representations*. Oxford: Basil Blackwell, 1980.

Chung, Yu Sun, Deanna Barch, and Michael Strube. "A Meta-Analysis of Mentalizing Impairments in Adults with Schizophrenia and Autism Spectrum Disorder." *Schizophrenia Bulletin* 40.33 (2014): 602–16.

Cialdini, Robert B., and Noah J. Goldstein. "Social Influence: Compliance and Conformity." *Annual Review of Psychology* 55 (2004): 591–621.

Clark, Kelly James. "Atheism and Analytic Thinking." In *The Science and Religion Dialogue: Past and Future*, edited by Michael Welker, 245–56. New York: Peter Lang, 2014.

———. "Atheism, Inference, and Intuition." In *Advances in Religion, Cognitive Science, and Experimental Philosophy*, edited by Helen De Cruz and Ryan Nichols, 103–18. London: Bloomsbury Academic, 2017.

———. "How Real People Believe: Reason and Belief in God." In *Science and Religion in Dialogue*, 2 vols., edited by Melville Y. Stewart, 1:481–99. Malden, MA: Wiley-Blackwell, 2010.

———. "Is Theism a Scientific Hypothesis? Reply to Maarten Boudry." *Reports of the National Centre for Science Education* 35.5, 5.2 (September–October 2015).

———. "Rappin' Religion's Solution to the Puzzle of Human Cooperation." *Huffington Post*, November 11, 2015.

———. "Reformed Epistemology and the Cognitive Science of Religion." In *Science and Religion in Dialogue*, 2 vols., edited by Melville Y. Stewart, 1:500–513. Malden, MA: Wiley-Blackwell, 2010.

———. *Return to Reason*. Grand Rapids: Eerdmans, 1990.

———. "Spirituality Wired: The Science of the Mind and the Rationality of Belief and Unbelief." *Revista Brasileira de Filosofia da Religião* 3.1 (2016): 9–35.

Clark, Kelly James, and Justin Barrett. "Reformed Epistemology and the Cognitive Science of Religion." *Faith and Philosophy* 27.2 (2010): 174–89.

———. "Reidian Religious Epistemology and the Cognitive Science of Religion." *Journal of the American Academy of Religion* 79.3 (2011): 639–75.

Clark, Kelly James, and Dani Rabinowitz. "Knowledge and the Cognitive Science of Religion." *European Journal of Philosophy of Religion* 3.1 (2011): 67–81.

Clark, Kelly J., and Ingela Visuri. "Autism and the Panoply of Religious Belief, Disbelief and Experience." In *The Neurology of Religion*, edited by A. Coles and J. Collicutt. Cambridge: Cambridge University Press, forthcoming.

Clifford, William K. *Lectures and Essays*, edited by Leslie Stephen and Frederick Pollock. London: Macmillan, 1886.

Conee, Earl, and Richard Feldman. "Evidentialism." *Philosophical Studies* 48 (1985): 15–34.

Crockett, Molly J., Luke Clark, Marc D. Hauser, and Trevor W. Robbins. "Serotonin Selectively Influences Moral Judgment and Behavior through Effects on Harm Aversion." *Proceedings of the National Academy of Sciences of the United States of America* 107 (2010): 17433–38.

Darwin, Charles R. *The Descent of Man*. London: J. Murray, 1874.

Dawkins, Richard. *The Blind Watchmaker: Why the Evidence of Evolution Reveals a Universe without Design*. New York: W. W. Norton, 1986.

———. *A Devil's Chaplain: Reflections on Hope, Lies, Science, and Love*. Boston: Houghton Mifflin, 2003.

———. *The God Delusion*. Boston: Houghton Mifflin, 2006.

De Cruz, Helen. "Cognitive Science of Religion and the Study of Theological Concepts." *Topoi* 33.2 (2013): 487–97.

Deeley, Quinton. "Cognitive Style, Spirituality, and Religious Understanding: The Case of Autism." *Journal of Religion, Disability and Health* 13.1 (2009): 77–82.

Dennett, Daniel C. *Breaking the Spell: Religion as a Natural Phenomenon*. New York: Viking, 2006.

Dickson, Lisa. "Race and Gender Differences in College Major Choice." *Annals of the American Academy of Political and Social Sciences* 627.1 (2010): 108–24.

Dougherty, Trent, ed. *Evidentialism and Its Discontents*. Oxford: Oxford University Press, 2011.

Doyen, Stéphane, Olivier Klein, Cora-Lise Pichon, and Axel Cleeremans. "Behavioral Priming: It's All in the Mind, but Whose Mind?" *PLOS One* 7.1 (2012).

Dubin, Nick, and Janet E. Graetz. "Through a Different Lens: Spirituality in the Lives of Individuals with Asperger's Syndrome." *Journal of Religion, Disability and Health* 13.1 (2009): 29–39.

Elqayam, Shira, and Jonathan Evans. "Subtracting 'Ought' from 'Is': Descriptivism versus Normativism in the Study of Human Thinking." *Behavioral and Brain Sciences* 34 (2011): 233–90.

Evans, C. Stephen. *Natural Signs and Knowledge of God: A New Look at Theistic Arguments*. New York: Oxford University Press, 2012.

Feldman, Fred. *Epistemology*. Upper Saddle River, NJ: Pearson, 2002.

Frederick, Shane. "Cognitive Reflection and Decision Making." *Journal of Economic Perspectives* 19 (2005): 25–42.

Frith, Uta, John Morton, and Alan M. Leslie. "The Cognitive Basis of a Biological Disorder: Autism." *Trends in Neurosciences* 14 (1991): 433–38.

Fumerton, Richard. *Epistemology*. Malden, MA: Wiley-Blackwell, 2006.

Gervais, Will M., and Ara Norenzayan. "Analytic Thinking Promotes Religious Disbelief." *Science* 336 (2012): 493–96.

Ghiselin, Michael. *The Economy of Nature and the Evolution of Sex.* Berkeley: University of California Press, 1974.

Glenn, Andrea L., Koleva Spassena, Ravi Iyer, Jesse Graham, and Peter H. Ditto. "Moral Identity in Psychopathy." *Judgment and Decision Making* 5 (2010): 497–505.

Goldman, Samuel. "Why Isn't My Professor Conservative?" *American Conservative*, January 7, 2016, http://www.theamericanconservative.com/articles/why-isnt-my-professor-conservative/.

Gould, Stephen Jay, and Richard C. Lewontin, "The Spandrels of San Marco and the Panglossian Paradigm: A Critique of the Adaptationist Programme." *Proceedings of the Royal Society B: Biological Sciences* 205 (1979): 581–98.

Grandin, Temple. *Thinking in Pictures: My Life with Autism.* Expanded ed. New York: Vintage Books, 2006.

Granqvist, Pehr, Mats Fredrikson, Patrik Unge, Andrea Hagenfeldt, Sven Valind, Dan Larhammar, and Marcus Larsson. "Sensed Presence and Mystical Experiences Are Predicted by Suggestibility, Not by the Application of Transcranial Weak Complex Magnetic Fields." *Neuroscience Letters* 379.1 (2005): 1–6.

Greco, John. "How to Reid Moore." *Philosophical Quarterly* 52.209 (2002): 544–63.

———. *Putting Skeptics in Their Place: The Nature of Skeptical Arguments and Their Role in Philosophical Inquiry.* Cambridge: Cambridge University Press, 2000.

Griffiths, Paul E., and John S. Wilkins. "Evolutionary Debunking Arguments in Three Domains: Fact, Value, and Religion." In *A New Science of Religion*, edited by Greg Dawes and James Maclaurin, 133–46. New York: Routledge, 2012.

Guthrie, Stewart. "A Cognitive Theory of Religion." *Current Anthropology* 21.2 (1980): 181–203.

———. *Faces in the Clouds.* Oxford: Oxford University Press, 1993.

Haidt, Jonathan. *The Righteous Mind: Why Good People Are Divided by Politics and Religion.* New York: Pantheon Books, 2012.

Hale, Frederick. "Religious Disbelief and Intelligence: The Failure of a Contemporary Attempt to Correlate National Mean IQs and Rates of Atheism." *Journal for the Study of Religion* 24.1 (2011): 37–53.

Hamer, Dean. *The God Gene: How Faith Is Hardwired into Our Genes.* New York: Doubleday, 2004.

Happé, Francesca G. E. "An Advanced Test of Theory of Mind: Understanding of Story Characters' Thoughts and Feelings by Able Autistic, Mentally Handicapped and Normal Children and Adults." *Journal of Autism and Developmental Disorders* 24.2 (1994): 129–54.

———. "The Role of Age and Verbal-Ability in the Theory of Mind Task-Performance of Subjects with Autism." *Child Development* 66.3 (1995): 843–55.

Harris, Sam. "This Must Be Heaven." *The Blog*, October 12, 2012, http://www.samharris.org/blog/item/this-must-be-heaven.

Haselton, Martie G. "The Sexual Overperception Bias: Evidence of a Systematic Bias in Men from a Survey of Naturally Occurring Events." *Journal of Research in Personality* 37 (2003): 43–47.

Haselton, Martie G., and David Buss. "Error Management Theory: A New Perspective on Biases in Cross-Sex Mind Reading." *Journal of Personality and Social Psychology* 78 (2000): 81–91.

Haselton, Martie G., and Daniel Nettle. "The Paranoid Optimist: An Integrative Evolutionary Model of Cognitive Biases." *Personality and Social Psychology Review* 10 (2006): 47–66.

Hegel, Georg Wilhelm Friedrich. *Encyclopaedia of the Philosophical Sciences*. 3rd ed. Translated by William Wallace. 1830.

Henrich, Joseph. "The Evolution of Costly Displays, Cooperation, and Religion: Credibility Enhancing Displays and Their Implications for Cultural Evolution." *Evolution and Human Behavior* 30 (2009): 244–60.

Henrich, Joseph, Steven J. Heine, and Ara Norenzayan. "The Weirdest People in the World?" *Behavioral and Brain Sciences* 33 (2010): 61–135.

Hercz, Robert. "The God Helmet." http://www.skeptic.ca/Persinger.htm.

Heywood, Bethany T. "Meant to Be: How Religious Beliefs, Cultural Religiosity, and Impaired Theory of Mind Affect the Implicit Bias to Think Teleologically." PhD diss., Queens University Belfast, 2010.

Heywood, Bethany T., and Jesse M. Bering. "'Meant To Be': How Religious Beliefs and Cultural Religiosity Affect the Implicit Bias to Think Teleologically." *Religion, Brain and Behavior* 4 (2013): 183–201.

Hill, Elisabeth, Sylvie Berthoz, and Uta Frith. "Cognitive Processing of Own Emotions in Individuals with Autistic Spectrum Disorder and in Their Relatives." *Journal of Autism and Developmental Disorders* 34 (2004): 229–35.

Hirschfeld, Lawrence A., and Susan A. Gelman, eds. *Mapping the Mind: Domain Specificity in Cognition and Culture*. Cambridge: Cambridge University Press, 1994.

Hodge, K. Mitch. "On Imagining the Afterlife." *Journal of Cognition and Culture* 11 (2011): 367–89.

Hodges, Sara D., Carissa A. Sharp, Nicholas J. S. Gibson, and Jessica M. Tipsord.

"Nearer My God to Thee: Self-God Overlap and Believers' Relationships with God." *Self and Identity* 12.3 (2012): 1–20.

Hodson, Gordon, and Michael A. Busseri. "Bright Minds and Dark Attitudes: Lower Cognitive Ability Predicts Greater Prejudice through Right-Wing Ideology and Low Intergroup Contact." *Psychological Science* 23 (2012): 187–95.

Hurlburt, Russell T., Francesca G. Happé, and Uta Frith. "Sampling the Form of Inner Experience in Three Adults with Asperger Syndrome." *Psychological Medicine* 24 (1994): 385–95.

Izuma, Keise, Kenji Matsumoto, Colin F. Camerer, and Ralph Adolphs. "Insensitivity to Social Reputation in Autism." *Proceedings of the National Academy of Sciences of the United States of America* 108.42 (2011): 17302–7.

James, William. *Pragmatism*. Indianapolis: Hackett, 1981.

———. *The Will to Believe and Other Essays in Popular Philosophy*. New York: Dover, 1956.

James, William, and Dominic Johnson. "God's Punishment and Public Goods." *Human Nature* 16.4 (2005): 410–46.

Johnson, Dominic, and Jesse Bering. "Hand of God, Mind of Man: Punishment and Cognition in the Evolution of Cooperation." In *The Believing Primate: Scientific, Philosophical, and Theological Reflections on the Origin of Religion*, edited by Michael J. Murray and Jeffrey Schloss, 26–44. Oxford: Oxford University Press, 2009.

Johnson, Dominic, and Oliver Krüger. "The Good of Wrath: Supernatural Punishment and the Evolution of Cooperation." *Political Theology* 5.2 (2004): 157–73.

Jolliffe, Therese, and Simon Baron-Cohen. "The Strange Stories Test: A Replication with High-Functioning Adults with Autism or Asperger Syndrome." *Journal of Autism and Developmental Disorders* 29.5 (1999): 395–406.

Jong, Jonathan. "Explaining Religion (Away?): Theism and the Cognitive Science of Religion." *Sophia* 52 (2012): 521–33.

———. "Implicit Measures in the Experimental Psychology of Religion." In *A New Science of Religion*, edited by Gregory W. Dawes and James McClaurin, 43–64. New York: Routledge, 2012.

Kahane, Guy. "Evolutionary Debunking Arguments." *Noûs* 45 (2011): 103–25.

Kahane, Guy, Jim A. C. Everett, Brian D. Earp, Miguel Farias, and Julian Savalescu. "'Utilitarian' Judgments in Sacrificial Moral Dilemmas Don't Reflect Impartial Concern for the Greater Good." *Cognition* 134 (2015): 193–209.

Kahneman, Daniel. *Thinking Fast and Slow*. New York: Farrar, Straus and Giroux, 2011.

Kang, Jerry, and Kristin Lane. "Seeing through Colorblindness: Implicit Bias and the Law." *UCLA Law Review* 58.2 (2010): 465–520.

Kapogiannis, Dimitrios, Aron K. Barbey, Michael Su, Giovanna Zamboni, Frank Krueger, and Jordan Grafman. "Cognitive and Neural Foundations of Religious Belief." *Proceedings of the National Academy of Sciences of the United States of America* 106.12 (2009): 4876–81.

Kelemen, Deborah. "Are Children 'Intuitive Theists'? Reasoning about Purpose and Design in Nature." *Psychological Science* 15.5 (2004): 295–301.

Kirkpatrick, Lee A. "Evolutionary Psychology as a Foundation for the Psychology of Religion." In *The Psychology of Religion: A Multidisciplinary Approach*, 2nd ed., edited by Raymond F. Paloutzian and Crystal L. Park, 118–37. New York: Guilford, 2013.

Koenigs, Michael, Michael Kruepke, Joshua Zeier, and Joseph P. Newman. "Utilitarian Moral Judgment in Psychopathy." *Social, Cognitive and Affective Neuroscience* 7.6 (2012): 708–14.

Kornblith, Hilary. *On Reflection.* Oxford: Oxford University Press, 2012.

Ladd, Kevin L., and Bernard Spilka. "Inward, Outward, and Upward: Cognitive Aspects of Prayer." *Journal for the Scientific Study of Religion* 41.3 (2002): 475–84.

Laird, Steven P., C. R. Snyder, Michael A. Rapoff, and Sam Green. "Measuring Private Prayer: Development, Validation, and Clinical Application of the Multidimensional Prayer Inventory." *International Journal for the Psychology of Religion* 14.4 (2004): 251–72.

Lawson, E. Thomas, and Robert N. McCauley. *Rethinking Religion: Connecting Cognition and Culture.* Cambridge: Cambridge University Press, 1990.

Libet, B., C. A. Gleason, E. W. Wright, and D. K. Pearl. "Time of Conscious Intention to Act in Relation to Onset of Cerebral Activity (Readiness-Potential): The Unconscious Initiation of a Freely Voluntary Act." *Brain* 106 (1983): 623–42.

Lind, Sophie E. "Memory and the Self in Autism Spectrum Disorder: A Review and Theoretical Framework." *Autism* 14 (2010): 430–56.

Lycan, William G. *Judgment and Justification.* Cambridge: Cambridge University Press, 1988.

Lynn, Richard, John Harvey, and Helmuth Nyborg. "Average Intelligence Predicts Atheism Rates across 137 Nations." *Intelligence* 37 (2009): 11–15.

Machery, Edouard, Ron Mallon, Shaun Nichols, and Stephen P. Stich. "Semantics, Cross-Cultural Style." *Cognition* 92.3 (2004): B1–B12.

McCauley, Robert N. *Why Religion Is Natural and Science Is Not.* Oxford: Oxford University Press, 2011.

McGinn, Colin. *The Problems of Philosophy: The Limits of Inquiry*. Hoboken, NJ: Wiley-Blackwell, 1993.

McNamara, Patrick, and Paul M. Butler. "The Neuropsychology of Religious Experience." In *The Psychology of Religion: A Multidisciplinary Approach*, 2nd ed., edited by Raymond F. Paloutzian and Crystal L. Park, 215–33. New York: Guilford, 2013.

Mercier, Hugo, and Dan Sperber, "Why Do Humans Reason? Arguments for an Argumentative Theory." *Behavioral and Brain Sciences* 34.2 (2011): 57–74.

Mitchell, Jason P. "Activity in Right Temporo-parietal Junction Is Not Selective for Theory-of-Mind." *Cerebral Cortex* 18.2 (2008): 262–71.

Morewedge, Carey K., and Michael Clear. "Anthropomorphic God Concepts Engender Moral Judgment." *Social Cognition* 26.2 (2008): 182–89.

Murray, Michael J. "Scientific Explanations of Religion and the Justification of Religious Belief." In *The Believing Primate: Scientific, Philosophical, and Theological Reflections on the Origin of Religion*, edited by Michael J. Murray and Jeffrey Schloss, 168–78. Oxford: Oxford University Press, 2009.

Murray, Michael J., and Andrew Goldberg. "Evolutionary Accounts of Religion: Explaining and Explaining." In *The Believing Primate: Scientific Philosophical and Theological Reflections on the Origin of Religion*, edited by Michael J. Murray and Jeffrey Schloss, 179–99. Oxford: Oxford University Press, 2009.

Närhi, Jani. "Beautiful Reflections: The Cognitive and Evolutionary Foundations of Paradise Representations." *Method and Theory in the Study of Religion* 20 (2008): 339–65.

Nichols, Ryan, and Paul Draper. "Diagnosing Cognitive Biases in Philosophy of Religion." *The Monist* 96.3 (2013): 420–46.

Nichols, Shaun, and Joshua Knobe. "Moral Responsibility and Determinism: The Cognitive Science of Folk Intuitions." *Noûs* 41.4 (2007): 663–85.

Nichols, Shaun, Stephen Stich, and Jonathan M. Weinberg. "Metaskepticism: Meditations in Ethno-Epistemology." In *The Skeptics*, edited by Stephen Luper, 227–47. Aldershot: Ashgate, 2003.

Nickerson, Raymond. "Confirmation Bias: A Ubiquitous Phenomenon in Many Guises." *Review of General Psychology* 2 (1998): 175–220.

Nietzsche, Friedrich. *The Gay Science: With a Prelude in Rhymes and an Appendix of Songs*. New York: Vintage, 2010.

Norenzayan, Ara. *Big Gods: How Religion Transformed Cooperation and Conflict*. Princeton: Princeton University Press, 2013.

Norenzayan, Ara, and Will M. Gervais. "The Origins of Religious Disbelief." *Trends in Cognitive Sciences* 17.1 (2013): 20–25.

Norenzayan, Ara, Will M. Gervais, and Kali H. Trzesniewski. "Mentalizing Deficits Constrain Belief in a Personal God." *PLOS One* 7.5 (2012).

Norenzayan, Ara, and Azim F. Shariff. "The Origin and Evolution of Religious Prosociality." *Science* 322 (2008): 58-62.

Norenzayan, Ara, Azim F. Shariff, Will M. Gervais, Aiyana K. Willard, Rita A. McNamara, Edward Slingerland, and Joseph Henrich. "The Cultural Evolution of Prosocial Religions." *Behavioral and Brain Sciences* 39 (2014): 1-86.

Norris, Pippa, and Ronald Inglehart. *Sacred and Secular: Religion and Politics Worldwide*. Cambridge: Cambridge University Press, 2004.

O'Hare, Anne E., Lynne Bremner, Marysia Nash, Francesca Happé, and Luisa M. Pettigrew. "A Clinical Assessment Tool for Advanced Theory of Mind Performance in 5 to 12 Year Olds." *Journal of Autism and Developmental Disorders* 39.6 (2009): 916-28.

Ozonoff, Sally, Bruce F. Pennington, and Sally Rogers. "Executive Function Deficits in High-Functioning Autistic Children: Relationship to Theory of Mind." *Journal of Child Psychology and Psychiatry* 32 (1991): 1081-1106.

Ozonoff, Sally, Sally Rogers, and Bruce F. Pennington. "Asperger's Syndrome: Evidence of an Empirical Distinction from High-Functioning Autism." *Journal of Child Psychiatry and Psychology* 32 (1991): 1107-22.

Pennycock, Gordon, James Allan Cheyne, Paul Seli, Derek J. Koehler, and Jonathan A. Fugelsang. "Analytic Cognitive Style Predicts Religious and Paranormal Belief." *Cognition* 123.3 (2012): 335-46.

Perner, Josef, and Heinz Wimmer. "'John Thinks That Mary Thinks That . . .': Attribution of Second-Order Beliefs by 5- to 10-year-old Children." *Journal of Experimental Child Psychology* 39 (1985): 437-71.

Piazza, Jared, Jesse M. Bering, and Gordon Ingram. "'Princess Alice Is Watching You': Children's Belief in an Invisible Person Inhibits Cheating." *Journal of Experimental Child Psychology* 109.3 (2011): 311-20.

Plantinga, Alvin. *Warrant: The Current Debate*. Oxford: Oxford University Press, 1993.

———. *Warrant and Proper Function*. Oxford: Oxford University Press, 1993.

———. *Warranted Christian Belief*. Oxford: Oxford University Press, 2000.

Plantinga, Alvin, and Nicholas Wolterstorff, eds. *Faith and Rationality: Reason and Belief in God*. Notre Dame: University of Notre Dame Press, 1984.

Poldrack, Russell A. "Can Cognitive Processes Be Inferred from Neuroimaging Data?" *Trends in Cognitive Sciences* 10.2 (2006): 59-63.

Premack, David, and Guy Woodruff. "Does the Chimpanzee Have a Theory of Mind?" *Behavioral and Brain Sciences* 4 (1978): 515-26.

Prinz, Jesse J. "Against Moral Nativism." In *Stich and His Critics*, edited by Michael Bishop and Dominic Murphy, 167–89. Malden, MA: Wiley-Blackwell, 2009.

Pyysiäinen, Ilkka. *Supernatural Agents: Why We Believe in Souls, Gods, and Buddhas*. Oxford: Oxford University Press, 2009.

Quine, Willard Van Orman. *Ontological Relativity and Other Essays*. New York: Columbia University Press, 1969.

Raver, Anne. "Qualities of an Animal Scientist: Cow's Eye View and Autism." *New York Times*, August 5, 1997. https://www.nytimes.com/1997/08/05/science/qualities-of-an-animal-scientist-cow-s-eye-view-and-autism.html.

Read, Max. "*Newsweek* Cover Story or Internet Posting about Drugs? A Quiz." *Gawker*, October 8, 2012. http://gawker.com/5949892/newsweek-cover-story-or-internet-posting-about-dugs-a-quiz.

Reddish, Paul, Penny Tok, and Radek Kundt. "Religious Cognition and Behaviour in Autism: The Role of Mentalizing." *International Journal for the Psychology of Religion* (2015): 95–112.

Reid, Thomas. *Essays on the Active Powers of the Human Mind.* 1788.

———. *Essays on the Intellectual Powers of Man.* 1785.

———. *An Inquiry into the Human Mind on the Principles of Common Sense.* 1764.

Richell, Rebecca A., Derek G. V. Mitchell, C. Newman, A. Leonard, Simon Baron-Cohen, and R. James R. Blair. "Theory of Mind and Psychopathy: Can Psychopathic Individuals Read the 'Language of the Eyes'?" *Neuropsychologia* 41.5 (2003): 523–26.

Rosenthal, Robert. *Experimenter Effects in Behavioral Research*. New York: Appleton-Century-Crofts, 1966.

Sacks, Oliver. "Seeing God in the Third Millennium." *Atlantic Monthly*, December 2012. https://www.theatlantic.com/health/archive/2012/12/seeing-god-in-the-third-millennium/266134/.

Sahraie, Arash, Paul B. Hibbard, Ceri T. Trevethan, Kay L. Ritchie, Lawrence Weiskrantz. "Consciousness of the First Order in Blindsight." *Proceedings of the National Academy of Sciences of the United States of America* 49.107 (December 2010): 21217–22; DOI:10.1073/pnas.1015652107.

Samson, Dana, Ian A. Apperly, Jason J. Braithwaite, Benjamin J. Andrews, and Sarah E. Bodley Scott. "Seeing It Their Way: Evidence for Rapid and Involuntary Computation of What Other People See." *Journal of Experimental Psychology: Human Perception and Performance* 36 (2010): 1255–66.

Saxe, Rebecca, and Anna Wexler. "Making Sense of Another Mind: The Role of the Right Temporo-parietal Junction." *Neuropsychologia* 43.10 (2005): 1391–99.

Schaap-Jonker, Hanneke, Bram Sizoo, Jannine van Schothorst–van Roekeld, and Jozef J. Corveleyn. "Autism Spectrum Disorders and the Image of God as a Core Aspect of Religiousness." *International Journal for the Psychology of Religion* 23.2 (2012): 145–60.

Schjoedt, Uffe, Hans Stødkilde-Jørgensen, Armin W. Geertz, and Andreas Roepstorff. "Highly Religious Participants Recruit Areas of Social Cognition in Personal Prayer." *Social Cognitive and Affective Neuroscience* 4.2 (2009): 199–207.

Shafir, Eldar, ed., *The Behavioral Foundations of Public Policy*. Princeton: Princeton University Press, 2012.

Shenhav, Amitai, David G. Rand, and Joshua D. Greene. "Divine Intuition: Cognitive Style Influences Belief in God." *Journal of Experimental Psychology: General* 141 (2012): 423–28.

Sherif, Muzafer. "An Experimental Approach to the Study of Attitudes." *Sociometry* 1.1 (1937): 90–98.

———. *The Psychology of Social Norms*. New York: Harper, 1936.

———. *A Study of Some Social Factors in Perception*. New York: Columbia University Press, 1935.

Simmons, Joseph P., Leif D. Nelson, and Uri Simonsohn. "False-Positive Psychology: Undisclosed Flexibility in Data Collection and Analysis Allows Presenting Anything as Significant." *Psychological Science* 22.11 (2011): 1359–66.

Smith, Adam. "The Empathy Imbalance Hypothesis of Autism: A Theoretical Approach to Cognitive and Emotional Empathy in Autistic Development." *Psychological Record* 59.3 (2009): 489–510.

Sosis, Richard. "The Adaptationist-Byproduct Debate on the Evolution of Religion: Five Misunderstandings of the Adaptationist Program." *Journal of Cognition and Culture* 9.3 (2009): 315–32.

———. "Does Religion Promote Trust? The Role of Signaling, Reputation, and Punishment." *Interdisciplinary Journal of Research on Religion* 1 (2005): 1–30.

———. "Religion and Intragroup Cooperation: Preliminary Results of a Comparative Analysis of Utopian Communities." *Cross-Cultural Research* 34 (2000): 70–87.

———. "Why Aren't We All Hutterites? Costly Signaling Theory and Religious Behavior." *Human Nature* 14 (2003): 91–127.

Spelke, Elizabeth, and Katherine D. Kinzler. "Core Knowledge." *Developmental Science* 11 (2007): 89–96.

Sperber, Dan. "Intuitive and Reflected Beliefs." *Mind and Language* 12.1 (1997): 67–83.

Sperber, Dan, Francesco Cara, and Vittorio Girotto. "Relevance Theory Explains the Selection Task." *Cognition* 57 (1995): 31–95.

Stark, Rodney. *What Americans Really Believe.* Waco, TX: Baylor University Press, 2005.

Stenger, Victor. "Not Dead Experiences (NDEs)." *Huffington Post,* October 11, 2012. https://www.huffingtonpost.com/victor-stenger/not-dead-exper eirnces-nde_b_1957920.html.

Stoet, Gijsbert, and David C. Geary. "Sex Differences in Mathematics and Reading Achievement Are Inversely Related: Within- and Across-Nation Assessment of 10 Years of PISA Data." *PLOS One* (2013).

Swain, Stacey, Joshua Alexander, and Jonathan M. Weinberg. "The Instability of Philosophical Intuitions: Running Hot and Cold on True Temp." *Philosophy and Phenomenological Research* 76.1 (2008): 138–55.

Swinburne, Richard. *Epistemic Justification.* Oxford: Oxford University Press, 2001.

———. *The Existence of God.* Oxford: Oxford University Press, 2004.

Szalavitz, Maia. "Q&A: An Interview with Oliver Sacks." *Time: Health & Family,* October 27, 2010. http://healthland.time.com/2010/10/27/mind -reading-an-interview-with-oliver-sacks.

Tager-Flusberg, Helen. "Language and Understanding Minds: Connections in Autism." In *Understanding Other Minds: Perspectives from Developmental Cognitive Neuroscience,* 2nd ed., edited by Simon Baron-Cohen, Helen Tager-Flusberg, and Donald J. Cohen, 124–49. Oxford: Oxford University Press, 2000.

Tager-Flusberg, Helen, and Robert M. Joseph. "How Language Facilitates the Acquisition of False Belief Understanding in Children with Autism." In *Why Language Matters in Theory of Mind,* edited by Janet Wilde Astington and Jodie A. Baird, 298–318. Oxford: Oxford University Press, 2005.

Tammet, Daniel. *Born on a Blue Day: Inside the Extraordinary Mind of an Autistic Savant.* New York: Free Press, 2007.

Thurow, Joshua. "Does Cognitive Science Show Belief in God to Be Irrational? The Epistemic Consequences of the Cognitive Science of Religion." *International Journal for Philosophy of Religion* 74 (2013): 77–98.

Tok, Penny. "Inner Speech Use in Autism." PhD diss., Victoria University of Wellington, New Zealand, 2013.

Tremlin, Todd. "Divergent Religion: A Dual-Process Model of Religious Thought, Behavior, and Morphology." In *Mind and Religion: Psychological and Cognitive Foundations of Religiosity,* edited by Harvey Whitehouse and Robert N. McCauley, 69–82. Walnut Creek, CA: AltaMira, 2005.

———. *Minds and Gods: The Cognitive Foundations of Religion.* New York: Oxford University Press, 2006.

Trigg, Roger, and Justin L. Barrett, eds. *The Roots of Religion: Exploring the Cognitive Science of Religion*. Farnham, Surrey: Ashgate, 2014.

Varga, Alexandra L., and Kai Hamburger. "Beyond Type 1 vs. Type 2 Processing: The Tri-Dimensional Way." *Frontiers in Psychology* 5 (2014): 993.

Visuri, Ingela. "Could Everyone Talk to God? A Case Study on Asperger's Syndrome, Religion and Spirituality." *Journal of Religion, Health and Disability* 16.4 (2012): 352–78.

Wechsler, David. *Wechsler Adult Intelligence Scale (WAIS-III)*. 3rd ed. San Antonio: Psychological Corporation, 1997.

———. *Wechsler Intelligence Scale for Children (WISC-III)*. 3rd ed. San Antonio: Psychological Corporation, 1991.

Weinberg, Jonathan M., Shaun Nichols, and Stephen Stich. "Normativity and Epistemic Intuitions." *Philosophical Topics* 29 (2001): 429–60.

White, Sarah, Elisabeth L. Hill, Francesca Happe, and Uta Frith. "Revisiting the Strange Stories: Revealing Mentalizing Impairments in Autism." *Child Development* 80.4 (2009): 1097–1117.

Whitehouse, Harvey. *Modes of Religiosity: A Cognitive Theory of Religious Transmission*. Walnut Creek, CA: AltaMira, 2004.

Whitehouse, Harvey, and Robert N. McCauley. *Mind and Religion: Psychological and Cognitive Foundations of Religiosity*. Walnut Creek, CA: AltaMira, 2005.

Whittington, B. L., and S. J. Scher. "Prayer and Subjective Well-Being: An Examination of Six Different Types of Prayer." *International Journal for the Psychology of Religion* 20.1 (2010): 59–68.

Wiech, Katje, Guy Kahane, Nicholas Shackel, Miguel Farias, Julian Savulescu, and Irene Tracey. "Cold or Calculating? Reduced Activity in the Subgenual Cingulate Reflects Decreased Aversion to Harming in Counterintuitive Utilitarian Judgment." *Cognition* 126 (2013): 364–72.

Wielenberg, Erik. "Evolutionary Debunking Arguments in Religion and Morality." In *Explanation in Ethics and Mathematics*, edited by Uri Leibowitz and Neil Sinclair, 83–102. Oxford: Oxford University Press, 2016.

Wigger, J. Bradley, Katrina Paxson, and Lacey Ryan. "What Do Invisible Friends Know? Imaginary Companions, God, and Theory of Mind." *International Journal for the Psychology of Religion* 23 (2013): 2–14.

Williams, Donna. *Nobody Nowhere: The Extraordinary Autobiography of an Autistic*. New York: Times Books, 1992.

Williams, Michael. *Problems of Knowledge: A Critical Introduction to Epistemology*. Oxford: Oxford University Press, 2001.

Williamson, Timothy. *Knowledge and Its Limits*. Oxford: Oxford University Press, 2000.

Wimmer, Heinz, and Josef Perner. "Beliefs about Beliefs: Representation and Constraining Function of Wrong Beliefs in Young Children's Understanding of Deception." *Cognition* 13 (1983): 103-28.

Wolterstorff, Nicholas. *Reason within the Bounds of Religion.* 2nd ed. Grand Rapids: Eerdmans, 1984.

———. *Thomas Reid and the Story of Epistemology.* Cambridge: Cambridge University Press, 2001.

Wood, Jay. *Epistemology: Becoming Intellectually Virtuous.* Downers Grove, IL: InterVarsity, 1998.

Wykstra, Stephen J. "Toward a Sensible Evidentialism: On the Notion of 'Needing Evidence.'" In *Philosophy of Religion: Selected Readings*, 2nd ed., edited by William Rowe and William Wainwright, 426-37. San Diego: Harcourt Brace Jovanovich, 1989.

Xie, Yu, and Kimberlee A. Shauman. *Women in Science: Career Processes and Outcomes.* Cambridge, MA: Harvard University Press, 2003.

Zimmer, Carl. "Faith-Boosting Genes: A Search for the Genetic Basis of Spirituality." *Scientific American* 291.4 (October 2004).

Zinnbauer, Brian J., and Kenneth I. Pargament. "Religiousness and Spirituality." In *The Psychology of Religion: A Multidisciplinary Approach*, edited by Raymond F. Paloutzian and Crystal L. Park, 21-42. New York: Guilford, 2005.

Zuckerman, Phil. "Atheism: Contemporary Numbers and Patterns." In *The Cambridge Companion to Atheism*, edited by Michael Martin, 47-65. Cambridge: Cambridge University Press, 2006.

Index